Safety, Reliability and Risk Management:
an integrated approach

Safety, Reliability and Risk Management:

an integrated approach

Second edition

Sue Cox and Robin Tait

OXFORD AMSTERDAM BOSTON LONDON NEW YORK PARIS
SAN DIEGO SAN FRANCISCO SINGAPORE SYDNEY TOKYO

Butterworth-Heinemann
An imprint of Elsevier Science
Linacre House, Jordan Hill, Oxford OX2 8DP
200 Wheeler Road, Burlington, MA 01803

First published 1991
Reprinted 1993
Second edition 1998
Transferred to digital printing 2003

British Library Cataloguing in Publication Data
A catalogue record for this book is available from the British Library

Library of Congress Cataloguing in Publication Data
A catalogue record for this book is available from the Library of
Congress

ISBN 0 7506 4016 2

For information on all Butterworth-Heinemann
publications visit our website at www.bh.com

Printed and bound in Great Britain by Antony Rowe Ltd, Eastbourne

Contents

Preface   vii

1   Introduction   1

2   Reliability   9

3   Greater reliability   27

4   Failure to danger   45

5   Human reliability: quantitative and qualitative assessment   63

6   Data sources   90

7   Human factors in system design   99

8   Programmable electronic systems   127

9   Outcomes and consequences   139

10   Ionizing radiation   160

11   Harm and risk   169

12   Quantitative risk analysis: limitations and uses   189

13   Risk assessment and cognition: thinking about risk   202

14   Risk assessment in occupational health and safety   221

15   Risk management and communication: decision-making and risk   235

16   Safety management principles and practice   263

17   Some recent incidents and their implications   290

Postscript   316

Appendix   318

Index   321

Preface

This book has been designed to support students and practitioners in their understanding and practice of safety and reliability. It reflects the experience and expertise of the authors in the fields of health and safety and health and safety education. It not only discusses issues relating to plant and technology but acknowledges the importance of human factors in the design, implementation and management of related safe systems. The adoption of this broad view of safety and reliability is quite deliberate and is a reflection of an almost universal movement towards a human factors orientation in this area. This stems not only from what has been learnt from the practice of health and safety management and risk assessment and from fundamental research, but also from the lessons drawn from recent investigations of major accidents and disasters.

The authors thus advocate an integrated approach to safety, reliability and risk management. This involves bringing together efficient engineering systems and controls of plant and equipment (hardware), not only with efficient management systems and procedures (software) but also with a practical understanding of people (liveware) and a general knowledge of other human factor considerations. This approach is compatible with the current development of risk management theory both in relation to public decision making (for example, land-siting decisions), and to loss prevention strategies. The engineering and human factors approaches are not incompatible, nor are they alternatives, but can be easily integrated into a single coherent view of the issues within the context of general systems theory (see Chapter 1).

The early chapters of the book (Chapters 2 and 3) consider reliability. They explore the background to developments in reliability engineering and the design of high-integrity systems. The principles are illustrated by examples drawn from the nuclear, chemical and aviation industries.

Chapter 4 and Chapter 5 develop these principles in the design of 'safe' and reliable systems. First, we consider the design of plant and equipment and utilize the procedures of hazard analysis including Hazard and Operability Studies (HAZOP), Hazard Analysis (HAZAN) and Failure Modes and Effects Analysis (FMEA). Second, we develop similar techniques in our understanding of human reliability. Chapter 6 provides information on safety and reliability data which support this design.

Chapters 7 and 8 develop an insight into the importance of human factors and new technology in modern sociotechnical systems. This importance is evident at all stages, including design, commissioning, operation and maintenance.

However, despite the efforts of safety and reliability specialists, systems failures still occur. Chapters 9 and 10 describe the outcomes and consequences of such failures in the nuclear and chemical industries and introduce the need for the quantification of the risk of a particular outcome. They build on the probabilistic approach described in an earlier section. Chapter 11 looks in more detail at harm and risk while Chapter 12 examines the process of quantified risk analysis, its limitations and its uses.

The role of individual cognitions of specific hazards and risks are reviewed in Chapter 13. This chapter also considers the relevance of such perceptions in developing the acceptance criteria used in the process of risk assessment.

The use of risk assessment in occupational health and safety is discussed in Chapter 14 while Chapter 15, dealing with risk management and communication and with land-use planning, has been written by an expert on the subject – Judith Petts, Deputy Director of the Centre for Hazard and Risk Management at Loughborough University. Chapter 16 discusses the application of risk management theory to safety management.

Finally, Chapter 17 describes some major incidents and their implications. The authors have emphasized learning points that can be drawn from the incidents as further elaboration of the risk management process.

A number of keywords are used in the book and they are defined at the appropriate point in the text. They include:

1. probability;
2. reliability;
3. safety;
4. hazard;
5. risk and risk management;
6. human error probability;
7. quantified risk analysis;
8. quantified risk assessment;
9. probabilistic risk assessment.

The definitions used in the book are based, wherever possible, on those used in *Nomenclature for Hazard and Risk Assessment in the Process Industries*, published by the Institution of Chemical Engineers (1985).

Acknowledgements

We would like to thank Phill Birch, Tom Cox and Stuart Tyfield for helpful discussions in the first edition and Sandy Edwards and Alistair Cheyne for their assistance with the preparation of the typescript.

Chapter 1

Introduction

Safety and reliability

A close association between safety and reliability has existed since the earliest times. When early communities first used spears and other weapons to protect themselves against wild animals, a broken shaft or blade could cost a life. They therefore needed to learn which woods and metals could be relied on (and which could not) in order to ensure safety. Indeed, the significance of the reliability of weapon design has remained up to the present day and much of the pioneering work on technical reliability has been carried out by the military.

Until the advent of modern scientific theory, technical progress was made by a sophisticated process of trial and error. Engineers and designers learned not only by their own mistakes but also from other people's misfortunes. This process was quite successful, as evidenced by the rapid progress made by master builders in the design of the great twelfth- and thirteenth-century cathedrals and abbeys. Admittedly, there were many dramatic building collapses when attempts were made to build vaults too high or columns too slim, but the survival of so many of these magnificent buildings provides living evidence of the development of their builders' skill.

The relationship between safety and reliability was intensified at the time of the Industrial Revolution. New sources of power, using water or steam, not only gave great potential for the rapid development of manufacturing technology but also provided a terrible potential for death and injury when things went wrong. The demand for new machinery and factory premises thus increased. In designing the necessary machines and buildings it had become possible to make use of the growing body of scientific knowledge, although designers still leaned heavily on past experience.

Scientific developments at that time were along strongly deterministic lines; theories strove to provide an exact and unambiguous account of natural phenomena. Failure to produce such an account was invariably considered to be a limitation of the theory rather than a fundamental impossibility.

The approach to safety and reliability during the Industrial Revolution was along similar deterministic lines. Structures were designed to a predetermined load factor or factor of safety, this being the ratio of the load predicted to cause collapse of the structure to the normal operating load. The factor was made large enough to ensure that the structure operated safely even if significant corrosion were present. Fixing the magnitude of the load factor was a relatively arbitrary process and led at times to considerable argument amongst the practitioners. The Brooklyn suspension bridge (1883) in the United States and the Forth railway bridge (1889) in Scotland provide good examples of well engineered structures of the period. However, the development of safety was only *partly* driven by such developments. Other important factors were the increased wealth and humanity of society and the economic value of the workforce. (Sadly, throughout history, humanity has had less influence than economics.)

Statistics and probability

Towards the end of the nineteenth-century, the sciences had begun to make use of statistical and probabilistic techniques (for example, in gas kinetics and genetics). This probabilistic approach entered the safety and reliability field on a large scale as attempts were made to operate electronic and other delicate equipment under battle conditions in World War II. Application of the well-tried factor of safety was no longer able to provide a solution and under the harsh operating conditions encountered on board ship or in combat aircraft, reliability (or rather unreliability) became a major problem. A typical airborne radar for instance, would do well to operate continuously for one hour without failure. In these circumstances it was necessary to study the causes and effects of component breakdown in order to improve reliability.

Still further demands have been placed on control and electronic equipment in the last few decades. Civil aircraft, for example, have increased greatly in size and complexity; there is more to go wrong and, with much increased passenger capacity, more lives are at risk. In the chemical industry, chemical reactors are larger than in earlier years and frequently operate under conditions where parameters such as temperatures and pressures must be very closely controlled to prevent run-away reactions. An accident to such a reactor could cost a great deal as well as having serious environmental and safety implications. Similar considerations apply in the nuclear industry both to reactor operation and to the handling and reprocessing of nuclear fuel. This continuing need to improve reliability has necessitated consideration of all aspects of sociotechnical systems and includes human factors.

The human factor

The study of human factors in systems reliability, through the application of psychology, ergonomics and human factors engineering, has grown dramatically over the past two decades. Early researchers in this area (for example, Spearman, 1928), bemoaned the seeming lack of interest in human error by their fellow psychologists: 'crammed as psychological writings are, and must needs be, with allusions to errors in an incidental manner, they hardly ever arrive at considering these profoundly, or even systematically'.

The most obvious impetus for this interest has been a growing public concern over the terrible costs of human error: the Tenerife air disaster in 1977, Three Mile Island (1979), Bhopal (1984) with its horrendous loss of life and Chernobyl (1986) with its implications for the public image of the nuclear power generating industry. The reliability of technology discussed in earlier sections has assumed an even greater significance today as the potential consequences of unreliable systems have become greater.

An additional spur to the developments in our understanding of human error has come from theoretical and methodological developments within cognitive psychology. It has become increasingly apparent that in order to provide an adequate picture of control processes, psychologists must explain not only correct performance but the more predictable varieties of human error. Reason (1990), in his book *Human Error*, maps the development of cognitive science in this area. Similarly, interest has focused on how operators respond to their working environment. In particular, ergonomists and human factors engineers have developed a greater understanding of human task performance and the interactions between humans and complex systems. The type and degree of human participation, especially in 'high' consequence areas, has been a matter of increasing concern. The pioneering work of Jens Rasmussen in this area (Goodstein *et al.*, 1988) is particularly noteworthy and will be discussed in later chapters.

Empirical data on human performance and reliability have also developed over recent years (see Chapter 6) and have fed into human reliability analysis (HRA). Such developments have been intimately tied up with the fortunes and misfortunes of the nuclear power industry. Although HRA techniques are increasingly used in other fields (for example, the offshore oil industry and chemical industry), much of the development in methodology has been associated with nuclear plant processes. This may, in some way, be linked with the public concern over the safety of nuclear power generation; a concern heightened by Chernobyl. In June 1988 the industry's technical magazine, *Nuclear Engineering International*, reported the results of its annual world survey, which showed that 10 countries, mostly in Europe, had cancelled reactor orders. It may also be a consequence of the need to demonstrate in advance that their reactor designs meet stringent safety criteria. In the United Kingdom, for example, these are expressed as order-of-magnitude probabilities less than 1×10^{-6} per reactor year for a large uncontrolled release.

The techniques used in designing to the very high reliability and safety standards required are very similar in all these applications. It is the purpose of this book to describe these techniques, to show how they may be applied and to highlight their limitations.

The systems approach

In this study of safety and reliability a systems approach has been adopted in which engineering, management procedures and human factors have been fully integrated (HSE, 1985, 1987). The systems approach is based on the application of general systems theory (see, for example, Checkland, 1981).

A 'system' is defined in the *Oxford English Dictionary* as 'a whole composed of parts in an orderly arrangement according to some scheme or plan'. There is some implication here of function and integrated goals. In relation to safety and reliability, systems can be treated as interacting or interdependent sets of components forming a network for the purpose of fulfilling some safety objective. Safety and reliability determinations need to encompass the measurement and integration of these separate components of the system. Park (1987) has developed a method for determining systems reliability which integrates technical reliability with human reliability (see Chapter 5).

Functionally systems are separated by distinct boundaries from the environment in which they operate. They are dynamic and purposeful (they *do* things). They import 'things' across their boundaries such as energy, information or materials, transform these inputs inside the system and then export some form of output back across the boundaries.

The issue of 'boundaries' is important. The concept is often obvious and useful in relation to biological or mechanical systems where it may

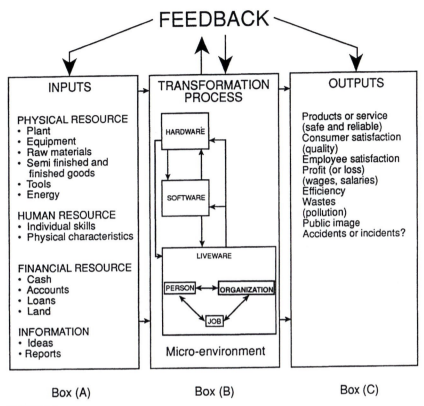

FIGURE 1.1
Systems model of an organization

have some physical basis. However, it is less obvious when discussing organizations and may be more conceptual than physical. For example, organizations operate under societal constraints, the values and opinions on the organizational systems of the neighbouring population may not match those of the organization itself. This can be illustrated schematically by a simplified inputs and outputs model (see Figure 1.1).

Box A represents the inputs into the system. In a typical manufacturing or service organization this would include the physical resources (for example plant, tools or energy), the human resource, financial resources and information. The transformation process (Box B) integrates the plant (hardware) and human resource (liveware), and utilizes the financial resources and information to develop organizational policy, procedures, rules and processes (software). The outputs of the system (Box C) are legion. They include a safe and reliable product or service, profits, social costs such as pollution and generate employee satisfaction and wages and salaries. If the transformation process is not designed and implemented in a safe and reliable way the system output *may* include an accident or incident.

The macro-environment (or operational environment) is an important influence on safety and reliability. This includes social, political, economic and legislative environments. It also includes the state of technology and its associated reliability.

Hazard and risk assessment

The notion of the system and its components completing various operations is best viewed in terms of probabilities of successful completion rather than simply as success or failure. Its assessment begins with an understanding of the various components of the system and a description of its function and goals.

From this description, safety and reliability practitioners can identify the potential sources of hazards and make an assessment of the associated risks (Cox and Tait, 1988). They make use of the techniques of quantified risk analysis (QRA) or probabilistic risk assessment (PRA) in this process. These are described in later chapters and are the foundations of risk management.

Risk management

Risk management is a technique which is increasingly used in organizations and by public bodies to increase safety and reliability and minimize losses. It involves the identification, evaluation and control of risks. *Risk identification* may be achieved by a multiplicity of techniques which are described in Chapters 4 and 5. *Risk evaluation* encompasses the measurement and assessment of risk. Implicit in the process is the need for sound decision making on the nature of potential socio-technical systems and their predicted reliability. The need for extra safety measures and guidance as to where they should be displayed are, in theory, the natural products of combined PRA/HRA studies. In an ideal world, good assessment should always drive effective error reduction.

Rasmussen and Pedersen (1984) have discussed the importance of PRA in the risk management process as a reference model to which risk management should aspire: 'The result of the PRA is a calculated risk which, if accepted, covers the "accepted risk". If not accepted, the design has to be modified until acceptance has been achieved'.

In practice, decisions on the acceptability of risk are dependent on other factors; these include social, economic, political and legislative concerns. A pragmatic evaluation often requires a balancing of risk reduction desirability against costs. This is illustrated in the discussions on land-use planning in Chapter 15.

The final stage in risk management is risk control. *Risk control* strategies may be classified into four main areas:

- Risk avoidance;
- Risk retention;
- Risk transfer;
- Risk reduction.

Risk avoidance involves a conscious decision on the part of the organization to avoid a particular risk by discontinuing the operation that is producing the risk. Risk retention may occur with or without knowledge:

1. With knowledge – a deliberate decision is made to retain the risk, maybe by self financing;
2. Without knowledge – occurs when risks have not been identified.

Risk transfer is the conscious transfer of risk to another organization, usually via insurance.

Risk reduction is the management of systems to reduce risks. It is the essence of this book and encompasses all the techniques, concepts and strategies that it describes in relation to technology, management systems and human factors. It thus concerns the engineers and technologists who design complex high-risk systems, those who develop the management procedures and, above all, those who manage and control the human factors.

References

Checkland, P. B. (1981) *Systems Thinking, Systems Practice*, John Wiley, Chichester.

Cox, S. and Tait, N. R. S. (1988), Quantified risk analysis and assessment. Successes and limitations, *The Safety Practitioner*, 9 May.

Goodstein, L. P., Andersen, H. B. and Olsen, S. E. (1988), *Tasks, Errors and Mental Models*, Taylor and Francis, London.

HSE (1985) *Deadly Maintenance: A Study of Fatal Accidents at Work*, HMSO, London.

HSE (1987) *Dangerous Maintenance: A Study of Maintenance Accidents in the Chemical Industry and How to Prevent Them*, HMSO, London.

Park, K. S. (1987) *Human Reliability Advances in Human Factors/Ergonomics*,

Elsevier, Amsterdam.

Rasmussen, J. and Pedersen, O. M. (1984) Human factors in probabilistic risk analysis and risk management, in *Operational Safety of Nuclear Power Plants*, Vol. 1, IAEA, Vienna.

Reason, J. (1990) *Human Error*, Cambridge University Press.

Spearman, C. (1928) The origin of error, *J. Gen. Psychol.*, **1**, 29.

Chapter 2

Reliability

Historical introduction

The early development of equipment reliability has been described by Shooman (1968) who reported that, whereas a typical destroyer in the United States Navy used around 60 electronic vacuum tubes in 1937, this number had risen to 3200 by 1952. This rapid expansion in the use of electronic equipment during and after World War II took place universally and involved all branches of the armed forces. In the United States Army, for example, Shooman (1968) reports that equipment was inoperative for as much as two-thirds or three-quarters of the time. In the United States Air Force, repair and maintenance over the lifetime of electronic equipment were costing ten times the capital cost of the equipment. In the United Kingdom similar difficulties were recorded. Dummer (1950) reported that airborne radar sets were only surviving about three hours' flying time, on average, without breakdown and that 600 000 radio valves were being used annually for military maintenance soon after the war. Dummer (1950) also quotes figures which demonstrate that significant losses can occur during transport and storage. In the very difficult conditions of the Far East campaign, it was estimated that about 60% of radar equipment was damaged during shipment and that half the surviving equipment deteriorated in storage (on arrival) to the point that it was not serviceable.

In conditions of war, electronic components are typically subjected to:

1. Impact;
2. Vibration;
3. Extremes of hot and cold;

4.   Humid and corrosive atmospheres;
5.   Atmospheric pressure cycling.

Soon after World War II, Nucci (1954) reported that 50–60% of electronic failures in one study were due to vacuum tube faults. He pointed out, however, that many failures were mechanical rather than electrical in origin. Great improvements in reliability were obtained in the 1950s when vacuum tubes were replaced by transistors. The reliability problems encountered at that time had obvious economic as well as military significance. This provided added impetus for improvement. There is much to learn from these early accounts of equipment failures and the methods used to improve reliability and to minimize down-time are still highly relevant today.

Design for reliability

The steps taken to improve reliability involve all stages in the design, construction and operation and are used for mechanical and electrical equipment as well as for electronic equipment.

Specification

The process starts at the equipment specification stage. Care is needed to specify the performance requirements and the range of operation. A power supply, for example, may be required to operate over the range 1–10 kv with a variability of less than 0.1% at ambient temperatures in the range − 10°C to 40°C. In practice, the situation is frequently very much more complex than this with performance being specified for a number of relevant variables. Operating conditions must be clearly specified, as component reliability is often very sensitive (for example, to changes in temperature or humidity). Some components, originally developed for use at normal atmospheric pressure have to be redesigned for use at high altitude, perhaps to withstand a pressure difference, or to compensate for reduced heat loss. Finally, of course, an overall reliability and perhaps an availability may well be specified. Such figures may be requested not only for complete equipment but for sub-assemblies or for critical components as well.

Design

The importance of minimizing mechanical failures has already been mentioned (Nucci, 1954). Mechanical design must provide adequate mechanical strength both for static and dynamic loads. Thus internal stresses in structures must be limited to acceptable levels, time dependent creep being taken into account where relevant. Where dynamic loads are present, the possibility of fatigue must be considered and designs must limit mechanical resonance and provide adequate protection against vibration and shock.

Components must be selected with great care. Frequently those produced to a general specification such as the Military Standards for United States electronic components or a British Standard, prove adequate. Where such components and specifications are not available, component development and test programmes may become necessary. Batch testing to ensure adequate quality control during component manufacture may often be necessary as well.

Prototypes

Frequently prototype units are constructed not necessarily using the final production techniques but to the full intended design. Such prototypes not only allow the general practicality of the design to be assessed but they also provide an opportunity for detailed prototype tests to be made. Such tests may well be repeated at a later stage on production units. Nucci (1954) reports that shock and vibration tests were instituted by the United States Navy in 1930 following serious equipment failures during trials on the cruiser Houston. Extensive tests were used during World War II both in the United States and the United Kingdom. In the United Kingdom, high temperature and high humidity cycles were employed. From 1946 the Standard K114 test added vibration, low temperature and low pressures (Dummer, 1950). Such tests are commonplace nowadays not only for military equipment but for a wide range of consumer goods and their components.

Construction and commissioning

In order to achieve the highest standards during construction, close attention must be paid to quality control. Detailed quality and performance records are required and formal certification is frequently requested. Many manufacturing and assembly processes demand high standards of cleanliness (for example, in integrated circuit production and in the assembly of high vacuum equipment). Adequate packaging is needed to ensure that equipment is not damaged in transit and suitable storage must be provided to prevent deterioration on arrival. Commissioning frequently requires careful planning, and the subsequent acceptance tests usually form part of the equipment specification.

Maintenance

Even if the equipment design is competent and the equipment is operated strictly within the limitations set in the specification, reliability will still depend strongly on the standards of maintenance and on the maintenance regime undertaken. Choice of regime will depend on a number of factors. If the costs of replacement parts are high and the disruption caused by breakdown is not great, breakdown maintenance may be appropriate. In this case, normal maintenance activities such as adjustment, servicing, running maintenance, minor repair or overhaul are undertaken only when breakdown occurs. On the other hand preventive maintenance undertaken to a predetermined schedule is designed to minimize disruption due to breakdown. Under this regime, components are overhauled or replaced before the time at which wearout is predicted from known lifetime data.

Availability is greatly influenced by design. Carefully planned diagnostic features and the provision of test points can be of considerable help in tracing faults, while modular design can speed up replacement. Such modular design with automatic plug-in interconnection of modules was very successfully employed early in World War II in Germany. Similar techniques are in common use today. The use of microprocessors to provide monitoring and diagnostic facilities has also had an increasing influence in recent years. We will return to this topic in a later chapter.

Human performance is of prime importance if high standards of reliability are to be achieved. Thus, personnel must work within a clear management structure in which all involved are fully aware of their own and others' responsibilities, all are given adequate training and provided with good supervision. The morale of the work force is also important. High morale is essential if high standards of work are to be attained. Human factors involved in system design are dealt with in more detail in Chapter 7.

Reliability in series and parallel

Before taking the study of reliability any further, some precise definitions are required. We define reliability as the probability that an item will perform its function under stated conditions for a stated period of time.

Thus, reliability (R) is a probability and as such may be anywhere in the numerical range $R = 1$ (perfect reliability and zero probability of failure) to $R = 0$ (complete unreliability and 100 per cent probability of failure). Unreliability (F) will now be $F = 1 - R$.

In order to take into account the periods during which repair or replacement follows breakdown it is also necessary to define availability. Availability (A) is defined as the probability that an item will be available at any instant of time. Thus, irrespective of the frequency of breakdown, $A = 1$ if repair or replacement is instantaneous, admittedly an unlikely situation.

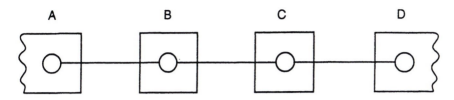

FIGURE 2.1
Series connection using three connectors

Series reliabilities

Using our probability-based definition of reliability, it is possible to predict how the overall reliability of a mechanism or equipment is influenced by the reliabilities of the components. This will depend on the form of the inter-connection between the components, the simplest being series connection (see Figure 2.1).

This is an example of an electrical connection being made between point A and point D by means of three wires, AB, BC and CD. The configuration will go open circuit if any one (or more) of the three wires is broken. Figure 2.1 is a literal example of series connection but the same effect, that is overall failure if one of a number of critical components fails, is encountered in much more complex situations where the components are not physically connected in series. Assuming that the reliabilities of the series components are mutually independent, Lusser's law of reliability (Lusser, 1950) states that the overall reliability is the product of the component reliabilities. Thus:

$$R = R_1 \times R_2 \times R_3 \times \ldots \times R_n$$

Lusser was the engineer in charge of the development of the V1 German pilotless bomber in World War II, and he encountered serious reliability problems during the very short period of time that was available to get the V1 operational. Lusser realized that the V1 had a great many 'series' type components. Of particular importance were the navigational system, the main engine valves, the gyro system and the one-way valve in the fuel line, but he listed about 100 components and sub-assemblies with measured failure rates of varying importance. Lusser's work has been reviewed by one of the current authors (see Tait, 1995).

The significance of the series reliability expression is demonstrated by making the simplifying assumption that all components have the same reliability. For 100 components, each of reliability $R_i = 0.99$, the overall reliability, using the formula, will be only $R = 0.37$. With 300 such components it would be reduced to $R = 0.05$. It is thus apparent that, in general, individual component reliability must be very much higher than the overall reliability required. The advantages of design simplicity, with the minimization in the number of components, are also apparent. For the more realistic case where all the reliabilities are different, Lusser realized that particular effort should be expended on the improvement of the least reliable components but that general improvement was frequently necessary in order to obtain acceptably high overall reliability.

Not all series systems obey Lusser's law of reliability although many electronics systems do seem to comply with it. For mechanical systems subjected to a broad range of loading strengths, the reliability tends to have a higher value. Thus $R = R_1$ where R_1 is one particular component reliability. This complex topic has been discussed by Carter (1986).

Parallel reliabilities

Greater complexity does not necessarily lead to decreased reliability, and this is the clue as to how reliability can be greatly improved in practice. In Figure 2.2 we see an example of parallel connection between A and B such that as long as one of the three connections has not failed we still have contact. The corresponding law for mutually independent components states that overall unreliability, F, is the product of the parallel component unreliabilities, thus:

$$F = F_1 \times F_2 \times F_3 \times ... \times F_n$$

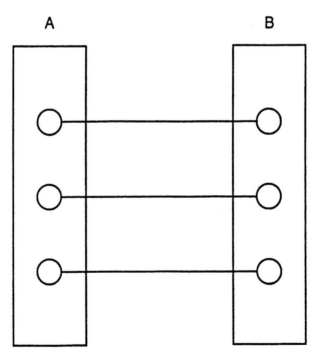

FIGURE 2.2
Parallel connection using three connectors

Again assuming for the moment that all component unreliabilities are equal, we can see how parallel operation helps. Taking component reliabilities of 0.6 or unreliabilities of 0.4 we see: for two in parallel $F = 0.4 \times 0.4 = 0.16$ and $R = 0.84$; for three in parallel $F = 0.4 \times 0.4 \times 0.4 = 0.064$ and $R = 0.936$. Thus we can produce equipment which is very much more reliable than its components. Note that the expression only applies where the component reliabilities are independent. Again the parallel connection might be literal, for example as in the provision of power to the electrical grid from two or more power stations, or it might not be so. The presence of a pilot in a normally pilotless aircraft could provide greatly improved effective reliability as he or she would be able in some cases to take 'parallel' compensatory action to counteract component failure.

Reliability prediction and design

By dividing an equipment design into sub-units having series or parallel interconnections it is frequently possible to make reasonably accurate reliability predictions without further complications although, as we will see later, more complex configurations can be dealt with. Such predictions allow equipment to be designed to a specified level of reliability and thus provide reliability control at the design stage. In order to control reliability at the operation stage, it is necessary to decide maintenance policy and this in its turn requires a knowledge of the time-dependence of component reliability.

The time variation of reliability

We concentrate on the normal and exponential lifetime distributions, although others are mentioned briefly.

The normal lifetime distribution

Figure 2.3 shows the results of a study made by Davis (1952) into the lifetime of 417 40-W light bulbs. All the bulbs were new and unused at the beginning of the test and Figure 2.3(a) is a histogram showing the times to failure plotted in 25-hour intervals. Note the considerable statistical fluctuations even with 417 samples. Superposed on the histogram is a smooth curve which is completely symmetrical about the central maximum. The correspondence between the two is quite good and we can see how it is possible to approximate closer and closer to such a curve by using more and more data and making the histogram steps smaller and smaller. The smooth curve is known as the normal curve. It can be expressed (see part 2 of the Appendix) in terms of two variables: the mean, which is also the operating time corresponding to the peak of the curve in Figure

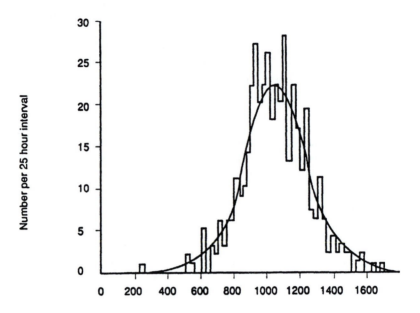

FIGURE 2.3 (a)
The lifetime histogram for 417 light bulbs with superposed normal curve

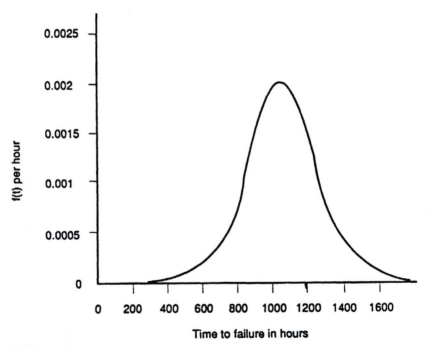

FIGURE 2.3 (b)
The corresponding failure density function f(t)

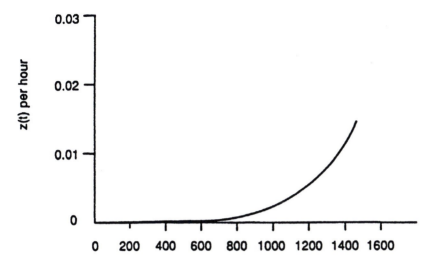

FIGURE 2.3 (c)
The hazard rate or failure rate z(t)

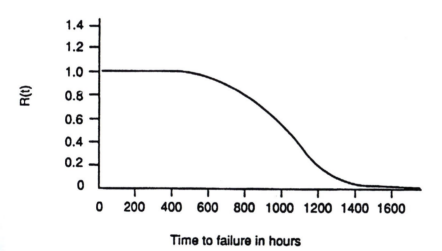

Time to failure in hours

FIGURE 2.3 (d)
The reliability R(t)

2.3(a), and the standard deviation, which is a measure of the width of the curve, representing the points at which the curve has dropped to 60.6% of its peak value. Examination of Figure 2.3(a) shows that the mean lifetime is 1050 hours and the standard deviation is about 200 hours, that is the 60.6% points are 200 hours on each side of the mean. The curve drops to 13.5% of the peak value at two standard deviations from the mean as can be confirmed approximately from Figure 2.3(a).

The failure density function

Davis' data are for 417 light bulbs. In order to make the information generally applicable to any number of these bulbs operating under similar conditions we divide by the total sample number. This will not alter the shape of the normal curve but whereas the peak value was 22 in Figure 2.3(a) it now becomes 22/417 = 0.0528. In Figure 2.3(b) we have also divided by 25 to give failure rates per hour rather than per 25-hour intervals. This curve, which represents the overall failure rate relative to the number existing at the start, is known as the failure density function $f(t)$.

The hazard rate

A second failure rate can be defined however, which is relative not to the total (initial) number, but to the number remaining at any subsequent time. Referring to the normal curve of Figure 2.3(a), we see that after 825 hours, 10 failures are taking place every 25 hours as there are after 1275 hours. But after 825 hours there are still about 30 bulbs left giving a failure rate of 10 out of 360 every 25 hours. After 1275 hours there are only 55 left and the failure rate is 10 out of 55 every 25 hours, so the likelihood of each individual bulb failing is now much larger. Thus if our interest is in the failure rate for each remaining bulb we divide not by the total number but by the remaining number.

The resulting curve, dividing also by 25 as previously, is given in Figure 2.3(c). The function $z(t)$ is known as the hazard rate or failure rate. It is seen to rise rapidly towards the end of the wear out process.

Both the failure density function and the hazard rate or failure rate are expressed as a number per unit time, per hour in our case. On the other hand, reliability $R(t)$ is a probability, that is a dimensionless number between zero and unity. $R(t)$ is defined as (the number remaining)/(total number). In the case of the light bulbs, all were working at $t = 0$, so $R(0) = 1$. By 1050 hours half had failed, so $R(1050) = 0.5$. The full curve is given in Figure 2.3(d).

The mathematical relationships connecting reliability, the failure density function and hazard rate are described in part 1 of the Appendix.

The exponential lifetime distribution

It is clear from Figure 2.3(a) that Davis observed no failures at less than 200 hours of operation even with a total sample as high as 417. The light bulbs then 'wore out' with a mean lifetime of 1050 hours. A completely different wearout behaviour is illustrated in Figure 2.4, again taken from Davis (1952). In this case 903 transmitter tubes were tested and all the samples failed within 1000 hours. The histogram interval

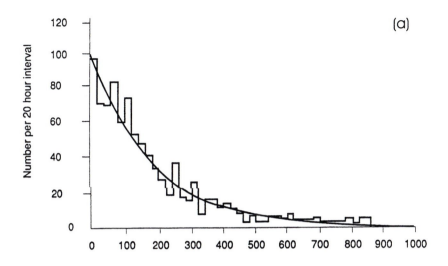

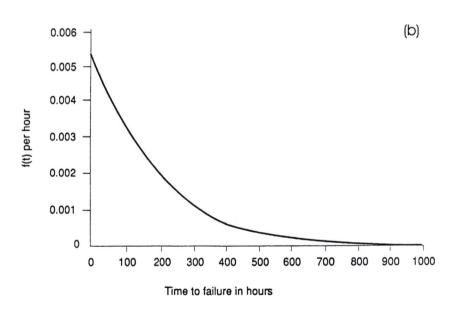

Time to failure in hours

FIGURE 2.4
*Lifetimes of 903 transmitter tubes. (a)   Lifetime histogram with superposed exponential
curve; (b)   Failure density function; (c)   Hazard rate; (d)   Reliability*

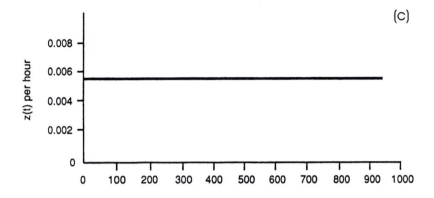

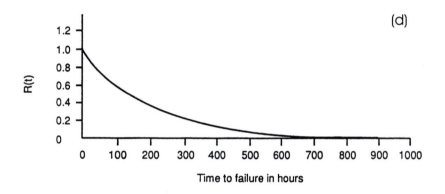

this time is 20 hours (Figure 2.4(a)) and a completely different failure pattern is now observed.

The failure density function $f(t)$ is found as previously by dividing the numbers failing per 20-hour interval by the total number, 903, and by the time interval. Thus the initial failure rate of 100 per 20 hours yields a failure density function value of $100/(903 \times 20) = 0.0056$ per hour. The full function which decreases steadily with time is seen in Figure 2.4(b).

The significance of this new failure mode is realized when the hazard rate is calculated (Figure 2.4(c)). This turns out to be a constant with time, indicating that the likelihood of each individual sample failing in a particular time interval does not depend on how long the sample has been running. The corresponding distribution is known as the exponential distribution. This can be expressed in terms of a single variable which is equal to the hazard rate and is also the reciprocal of the mean life of the samples.

Thus in the present case the mean life of the 903 tubes was 179 hours and the hazard rate is thus 1/179 = 0.0056 per hour. Because of the rapid initial rate of failure, the reliability falls fairly rapidly at first, but decreases more slowly as time passes (Figure 2.4 (d)). The curve is in fact of identical shape to the failure density function. More details of the exponential distribution are given in part 3 of the Appendix.

Multiple failure modes

The constant value of the hazard rate, independent of time, can result from occasional purely random variations of a mechanical load or temperature or pressure to such an extreme value that failure takes place. Similar breakdown characteristics are also observed under other conditions however. A constant hazard rate can be produced when more than one failure mode is present these having different failure rates which happen to combine to produce a constant overall rate. Such a situation is easily distinguished if the failure modes can be separately identified and shown to have non-constant hazard rates. A second situation in which non-random failures can produce a constant hazard rate is where repair or replacement with new components follows immediately on failure.

This is well illustrated in Figure 2.5, again based on Davis (1952), in which bus engines are studied. The first failures are seen to have an approximately normal time distribution but by the time the fourth failures are encountered an exponential distribution has been established.

The exponential distribution takes a particularly simple form and is relatively easy to handle in reliability calculations. It does apply in a wide range of situations, but very serious errors can be incurred if it is used under inappropriate conditions.

Replacement policy

It is possible to conceive of equipment exhibiting an initial random failure mode followed by a normal wearout peak. Scheduled maintenance would improve availability if it was designed to replace components towards the end of the random failure period before wearout failure had set in to a significant extent. If replacement took place too soon on the other hand, costs would rise but availability would not be significantly improved. Correct selection of replacement time obviously requires knowledge of the failure density function for the components involved.

Early failure

A third regime, during which early failures take place, is also frequently encountered. In this regime, hazard rate decreases with time. Early

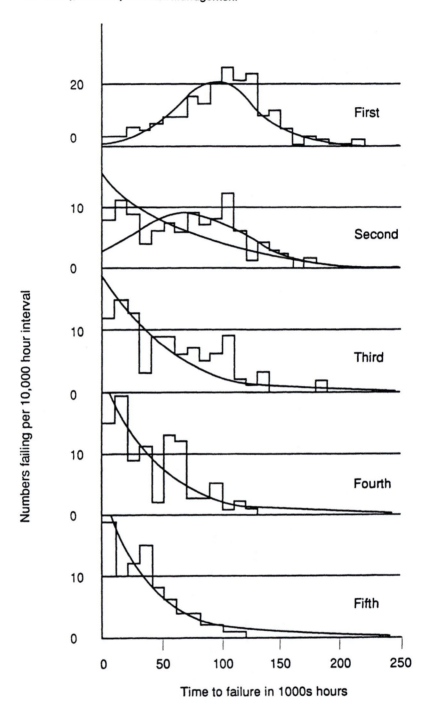

FIGURE 2.5
Histograms of the successive failures of bus engines. The first failure distribution is of normal form but by the fifth, the distribution has taken the exponential form

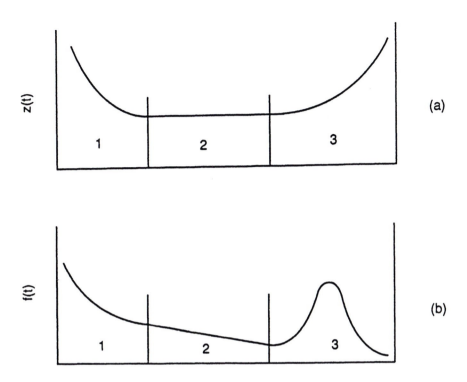

FIGURE 2.6
*The 'bathtub' curve. (a) Hazard rate z(t); and (b) failure density function f(t). Regions 1, 2 and
3 are respectively the early, constant and wearout failure regimes*

failure can be due to the relatively rapid breakdown of components
which are either faulty or substandard. This is often due to poor
quality control during manufacture. It can also be due to incorrect
installation procedures and can even be caused by poor maintenance;
in which case the equipment may never reach the random failure
regime before starting to encounter the increasing hazard rate due to
wearout. Early failure can frequently be greatly reduced by 'burning
in' components – running them for a while before installation in order
to eliminate weak components. This procedure is commonly employed
with electronics equipment.

A schematic drawing, with all three failure patterns present, is to be
found in Figure 2.6. In Figure 2.6(a) the hazard rate is shown, producing
the characteristic 'bathtub' curve. The corresponding failure density
function curve is reproduced in Figure 2.6(b). The situation here is
idealized. Many electronic components fail almost invariably in the
early or random regimes while some mechanical systems exhibit little
or no random failure.

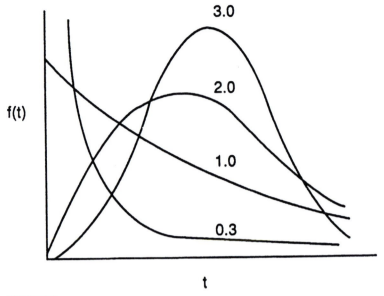

FIGURE 2.7
Failure density functions for various beta values of thge Weibull distribution

Other distributions

By measuring failure rates it is possible to distinguish between early, constant and wearout failure although the different breakdown modes must be examined separately to ensure that the constant region is a real exponential. It is, however, possible to discover a great more, particularly in the wearout region. In order to do this several other distribution functions are used. The Weibull distribution (Weibull, 1951) is a particularly useful one. Varying one of the Weibull function parameters, β, allows greatly varying distributions to be reproduced, Figure 2.7. For β less than 1, both $f(t)$ and $z(t)$ fall with increasing t as in the early failure regime. For $\beta = 1$, the exponential distribution is obtained while for $\beta = 2$ the hazard rate $z(t)$ increases linearly with time and the failure density function is like a normal peak but with a long 'tail' to the right of the peak. For $\beta = 3.4$, a close-to-normal distribution is produced. Weibull was able to obtain good fits to several populations (groups of objects) including the yield strengths of Bofors steel samples, $\beta = 2.9$, the size distribution of fly-ash, $\beta = 2.3$, and the strength of Indian cotton, a very asymmetric distribution with $\beta = 1.46$. Some of the populations studied turned out to have more than one component present, the components having different physical properties and different β values.

Another distribution function in common use is the Log-normal distribution which also has a 'tail' to the right of the peak. It is often used to fit distributions representing crack propagation, corrosion rates, bacterial attack and fatigue. Other specialized functions include the Gamma distribution, Extreme Value and Birnbaum–Saunders distributions (see Carter, 1986, or Lees, 1996, for example). Parameter fits obtained using these distributions can frequently give clues to the nature of the failure mechanism involved or of the previously unsuspected presence of more than one component as in Wiebull's paper. Such parametrization also allows predictions to be made of how failure rates would be expected to change with variation of design, thus allowing redesign with improved reliability.

Conclusions

This chapter has shown how reliability can be calculated for simple configurations of components in terms of the measured reliabilities of the components. It has illustrated this with reference to a number of examples. It has also investigated failure mechanisms and their statistical distributions. Chapter 3 will consider how this information can be used to produce equipment to the highest reliability standards.

Further reading

Billington, R. and Allen, R. N. (1983) *Reliability Evaluation of Engineering Systems*, Pitman, London.

BS 5760 (1979, 1981, 1982) *Reliability of Systems, Equipment and Components*, British Standards Institution, Parts 1, 2 and 3, London.

Carter, A. D. S. (1986), *Mechanical Reliability* (2nd ed.), Macmillan, Basingstoke.

O'Connor, P. D. T. (1984) *Practical Reliability Engineering*, John Wiley, Chichester.

Smith, D. J. (1985) *Reliability and Maintainability in Perspective*, Macmillan, Basingstoke.

References

Carter, A. D. S. (1986) *Mechanical Reliability* (2nd ed.), Macmillan, London.

Davis, D. J. (1952) An analysis of some failure data, *J. Am. Stat. Assoc.*, **47**, 113.

Dummer, G. W. A. (1950) A study of the factors affecting the reliability of radar equipments, *TRE Technical Note 89*.

Lees, F. P. (1996) *Loss Prevention in the Process Industries* (2nd ed.),

Butterworth-Heinemann, Oxford.

Lusser, R. (1950) A study of methods for achieving reliability in guided missiles, *NAMTC Technical Report No. 75.*

Nucci, E. J. (1954) The Navy Reliability Program and the Designer, *Proc. 1st Nat. Symp. on Quality Control and Reliability*, New York, November 1954, p. 56.

Shooman, M. L. (1968) *Probabilistic Reliability: An Engineering Approach*, McGraw-Hill, New York.

Tait, N. R. S. (1995) Robert Lusser and Lusser's law, *Safety and Reliability*, **15**(2), 15.

Weibull, W. (1951) A statistical distribution function of wide applicability, *J. Appl. Mech.* **18**, 293.

Chapter 3

Greater reliability

Reliability enhancement

Much of the pioneering study of reliability was associated with control and electronics equipment. Chapter 2 described how the early, random and wearout phases of component or unit failure were recognized and how some of the types of failure distribution associated with different failure processes were identified. Once the product rule for the combined reliability of components in a series configuration had been recognized, the importance of design simplicity became apparent. The product rule also focused attention on the importance of care in component selection and of quality control in securing better overall reliability. One way a particular component could be made to give more reliable performance was to derate it. For example, electrical and electronic components can be operated at reduced current or voltage, mechanical components at reduced stress level or torque. Another way of enhancing reliability was to employ parallel configurations of similar components as described in Chapter 2.

The development of new components, very often working on completely new principles, has also produced very great improvements in reliability. This is particularly evident with electronics equipment where the electronic vacuum tube was successively replaced by the transistor and then by the integrated circuit. Such improvements have been essential in order to obtain acceptable reliability in the very large electronic configurations in modern use in computers, telecommunications and RADAR equipment.

This chapter describes the methods used to further enhance reliability. It also provides examples of several high integrity systems.

High integrity systems

Technical developments have contributed greatly to improved reliability but are not sufficient on their own to provide the performance demanded of the highest integrity equipment. This is illustrated by the following reliability considerations. A mean time to failure for a surface to air missile of 30 minutes might well be considered satisfactory. Similarly a mean time between failures of 10 hours would probably be acceptable for a large land-based radar which has been designed for rapid fault diagnosis and repair. On the other hand, the United Kingdom Air Registration Board demands that the automatic landing systems on civil aircraft should operate such that there is less than one fatality in every 10 000 000 landings, as described by Lloyd and Tye (1982). These standards are necessary since failure of the automatic landing system on a large aircraft could lead to many hundreds of deaths, as well as massive financial losses and a loss of good name both for the manufacturer and the airline. Similar high reliability standards are demanded in a range of other situations where human safety is involved; for example, the emergency shut-down systems used in chemical and nuclear reactors. We designate these as high integrity systems.

Parallel redundancy

The single most important technique for generating high integrity systems is the employment of components in parallel configurations. Chapter 2 demonstrated how overall reliability can be greatly improved in this way. This use of more than one component in order to perform a function for which only a single component is strictly necessary is known as redundancy. Many simple examples of redundancy can be quoted:

1.   Frequently two diagonally opposite wheels on a car have their brakes controlled by one hydraulic line while the other two are controlled by a second independent one.
2.   Two-engined aeroplanes are normally designed so that they can, if necessary, fly using only one engine.
3.   The electrical supply network has many complex interconnecting links such that electrical supplies can be maintained even if several links are lost.

Many mechanical structures such as steel frame buildings and lattice girder bridges have far more struts and ties than are strictly necessary. Such designs are frequently employed in order to facilitate manufacture or construction, to reduce internal stresses or to increase the rigidity of the structure. A very simple example of mechanical redundancy is illustrated in Figure 3.1 which is based on an example given in *Reliability Technology* (Green and Bourne, 1972). The individual members are

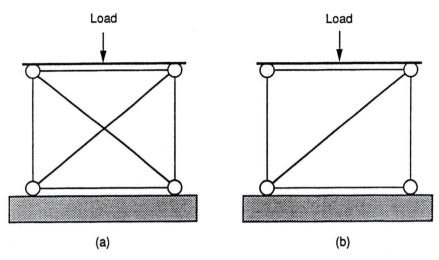

FIGURE 3.1
Redundancy in a simple mechanical structure with flexible joints. (a) Redundant; and (b) non- redundant

assumed to be perfectly rigid and the joints between them perfectly flexible. Structure (a) is redundant in that it retains its form if any single member fails. Structure (b) contains no redundancy; it will collapse if any single member fails.

We will now examine in more detail how parallel redundancy is employed in practice.

Mathematical expressions are quoted and used to calculate simple reliabilities as illustrative examples. The derivations of the expressions are to be found in the texts quoted at the end of the chapter.

It was shown in Chapter 2 that for parallel configurations the overall unreliability is equal to the product of the component unreliabilities. Thus for two parallel components:

$$F = F_1 \times F_2$$

while for three parallel components:

$$F = F_1 \times F_2 \times F_3$$

The two component case is illustrated in Figure 3.2(a). As we have already discussed, the two components run in parallel and the configuration continues to operate successfully as long as one component is still running. Taking the simple example where $F_1 = F_2 = 0.05$, that is, the reliabilities are $R_1 = R_2 = 0.95$, then:

$$F = F_1 \times F_2 = 0.05 \times 0.05 = 0.0025$$

and

$$R = 1 \times F = 0.9975$$

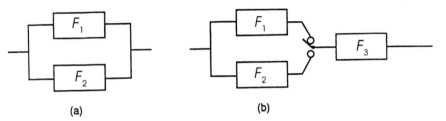

FIGURE 3.2
Components in parallel. (a) Simple parallel redundancy; (b) standby redundancy

Standby redundancy

In cases where a component fault or failure is revealed, that is, it is detected as soon as it occurs, then it is possible to use standby redundancy. Many simple examples of standby redundancy can be quoted, such as the battery operated emergency lighting which switches on automatically when the main electricity supply fails, the reserve players at a football match or the fine chain and safety pin which secures a valuable brooch in addition to the main fastening pin.

Standby redundancy is illustrated in Figure 3.2 (b), where the additional element F_3 is the failure probability for the switchover process. The overall unreliability is now the (Probability of main and standby elements both failing but switchover working) + (Probability of main system failing and switchover failing). Thus:

$$F = F_1F_2(1 \times F_3) + F_1F_3$$

Taking $F_1 = F_2 = F_3 = 0.05$
we find

$$F = 0.004875$$

and

$$R = 0.995125$$

Thus the reliability has reduced somewhat because of the unreliability of the switchover mechanism. In real cases, standby is used where restoration of normal operation by repair of component 1 is rapid and where standby operation with component 2 can be at a somewhat limited level. In this case component 2 may be relatively inexpensive and F_1 and F_2 may be quite different. Indeed, component 1 might be a composite of several parallel redundant subcomponents providing a very low failure probability as in our earlier example. The switchover mechanism performs a completely different function to the other components so will in general have a different unreliability. The assumption that $F_1 = F_2 = F_3$ is therefore an oversimplification.

Fail-safe design

So far, in studying parallel redundancy we have concentrated on the calculation of the probability of failure, that is the unreliability. Of course not all failure modes will have the same effect, and in general, high integrity circuits are designed wherever possible so that they fail to a safe condition. A simple example of fail-safe design is the 'dead mans handle' employed on teach pendants of robot systems, see Chapter 8. This must be held down before the robot can be moved. If the driver becomes incapacitated the pressure on the handle will be removed, bringing the robot to a rapid (and hopefully safe) halt. The problem with failure to safety in control and shutdown systems is that it leads to spurious interruption in the process or procedure which is under way. Although such interruption is 'safe', it can lead to other dangers, causing disruption to operating procedures, distracting the operating crew and bringing the system into disrepute.

Voting procedures

Such effects can be greatly reduced by employing three or more parallel redundant components, each designed to be failsafe, but then using a voting procedure. For three components, no action is taken unless at least two of them demand a shut down.

Thus in Figure 3.3(a) we have a single component with failure probability F_1. Since the system has been designed to be failsafe there is no hazard associated with failure but there is a probability F_1 of a spurious trip. In Figure 3.3(b) with its two out of three vote, the overall probability of spurious trip, assuming $F_1 = F_2 = F_3$ is the (Probability that all fail) + (Probability that any two fail while the third works).

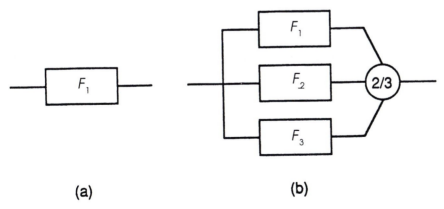

(a)   (b)

FIGURE 3.3
Comparison of (a) a single component with (b) two out of three voting

Hence:

$$F = F_1^3 + 3(1 - F_1)F_1^2.$$
$$= F_1^2(3 - 2F_1)$$

Taking, for example, $F_1 = 0.05$, $F = 0.00725$ compared with 0.05 for the single component. Still further improvement is obtained if three out of four logic is used. In practice, life is not as simple – many components have more than one failure mode such that they may, at times, fail to danger. However, the simple example shows how spurious trips can be greatly reduced by using majority voting techniques.

Fractional dead time

Emergency shutdown equipment will inevitably fail to danger in some circumstances. A hazard will then arise if the equipment is activated. Three quantities are involved in assessing such a situation:

1. *The hazard rate* – the number of occasions per unit time (per year for example) upon which the hazard is expected to arise;
2. *The demand rate* – the number of occasions per unit time upon which the shutdown equipment is called on to operate;
3. *The fractional dead time* – the fraction of the time during which the shutdown equipment is inoperative.

Thus:

Hazard rate = Demand rate × Fractional dead time

If the fractional dead time is 0.01 and the demand rate is two per year, the hazard rate is $2 \times 0.01 = 0.02$ per year. This means that the hazard may arise on average once in 50 years. Note that the fractional dead time is $(1 - A)$, where A is the availability as defined in Chapter 2. In cases where a fault is revealed, that is, it is detected as soon as it fails, the fractional dead time depends on the length of time required to replace or repair the component. Where the fault is unrevealed, the faulty condition will be undetected until a functional test is performed. We will assume that such tests are made at regular intervals of time T. Normally T is chosen to be much less that $1/f$, where f is the frequency at which the unrevealed faults take place on average. We can then normally assume that when a fault does occur it is equally likely to be at any time within the test interval T. Thus, on average, the fault will be undetected for time $T/2$ after a time interval of $1/f$, so the fractional dead time is $(T/2)/(1/f) = fT/2$.

For example, if the test interval is one week and the fault occurs on average once per year, say once in 50 weeks, then $f = 1/50 = 0.02$ per week and the fractional dead time is $fT/2 = 0.02 - 1/2 = 0.01$.

Note that f and T must use the same units of time, weeks in our example. In practice, our aim of keeping the fractional dead time low is best achieved by making f small, that is by making the system highly reliable, rather than by making T small and thus having relatively frequent tests. This is because in practice the testing procedure can in itself cause failure or add to dead time.

The methods used here to calculate fractional dead time and hazard rate contain a number of assumptions and approximations, but this simple treatment allows the principles to be established. Further information is to be found in Part 4 of the Appendix and in Lees (1996).

Complex configurations

In Chapter 2, both series and parallel configurations were introduced. In the present Chapter we have discussed how parallel systems can be used to produce reliabilities which are far superior to those of the separate components, and also to reduce spurious trips in systems which are fail-safe. These techniques are absolutely fundamental to high reliability design although in practice equipment frequently contains a complicated configuration of interconnected series and parallel elements. In some cases these reduce very easily to a simple configuration. For example, taking the arrangement of Figure 3.4(a) we can combine series elements A1 and A2 to give A with a reliability given by Lusser's law. B1 and B2 are combined similarly (Figure 3.4(b)). Next A and B are combined in parallel to give element W, as are C and D to give X using the law of parallel combination (Figure 3.4(c)). Similarly W and X are series-combined (Figure 3.4(d)) and Y and E parallel combined to give a single reliability value for element Z.

A similar approach is not possible in a configuration like that of Figure 3.5, but alternative techniques are available to calculate the overall reliability in such cases, and to calculate availabilities in situations where repairs or replacements take place, see for example Andrews and Moss (1993) or Lees (1996). Thus, using these various techniques, it is possible to predict the reliabilities and availabilities of complex equipment configurations, and hence by making improvements at the appropriate places to produce a design to any specified standard of reliability or availability.

Limits attainable on reliability

There must, of course, be a limit to the reliability or availability that can be reached. By using parallel redundancy we can, in principle, build

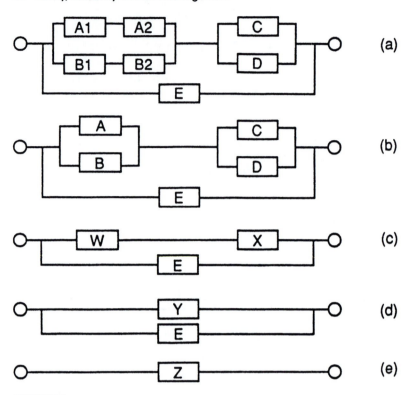

FIGURE 3.4
Simplification (a) to (e) by successive combination in series or parallel

up very high levels of reliability. For example, if one element has for a one year period an unreliability of 0.1, two in parallel will have 0.01, three in parallel 0.001 and four in parallel will have 0.0001. But the expression used to calculate parallel reliability is only valid if the component reliabilities are strictly independent of one another. Two very common effects can compromise this independence; common cause or common mode failure and cascade failure.

Common mode failure

Common mode failure results when a single factor, for example a loss of electrical power or a mechanical failure, simultaneously causes failure in two or more redundant components. Many simple examples can be quoted of common mode failure. For example, a well designed building may collapse due to inadequate foundations or a carefully organized and well trained Works Fire Team can be made completely ineffective if their means of intercommunication fails, or if their fire tender fails to start.

This type of failure can have a very serious effect on the reliability of high integrity systems and a great deal of effort goes into the elimination or at least minimization of such processes. We can model

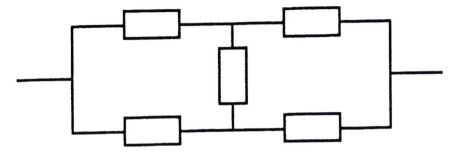

FIGURE 3.5
Example where simplification by combination in series or parallel is not possible

the presence of the common mode failure by assuming it to be in series with the redundant components. If the former has an unreliability of 1×10^{-3} and the latter of 1×10^{-4}, Lusser's Law gives a combined unreliability of 1.1×10^{-3}, close to that of the common mode failure and almost a factor of ten worse than that of the redundant system, clearly an unacceptable situation. The ratio of the common mode failure probability to the total failure probability is known as the β ratio. Thus in our example:

$$\beta = 1 \times 10^{-3}/1.1 \times 10^{-3} = 0.91$$

Note that β approaches 1.0 as common mode failure becomes more and more dominant, and 0 as common mode failure becomes negligible.

Some examples of types of common mode failure are to be found in Figure 3.6. In Figure 3.6(a) two parallel redundant components B_1 and B_2 are controlled or powered by A (A could be an electrical, hydraulic or mechanical actuator for example). If A fails, then both B_1 and B_2 are inactivated by common mode failure. In Figure 3.6(b) loss of control or power is made less probable by replacing A by parallel units A_1 and A_2. But these do act at a single point x so a break of linkage at x can still cause common mode failure. This configuration is frequently found where two independent hydraulic circuits activate a simple mechanical component.

In Figure 3.6(c) the situation is improved further by duplicating both A and B, but common mode failure can still take place in the return circuit (electrical or hydraulic for example) at y. This return circuit might be a single conductor as drawn in Figure 3.6(c) or it might be two separate ones which are physically close together and thus susceptible to common mode failure by fire, by chemical attack or by physical damage. In Figure 3.6(d) the return circuits have been made more independent by physical segregation. Edwards and Watson (1979) have given a detailed description of various real examples of common mode failure and they discuss how common mode failure can be minimized.

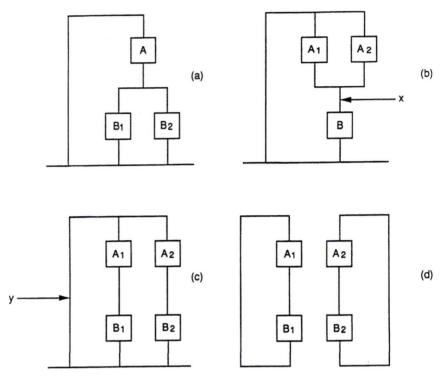

FIGURE 3.6
Reduction in common mode failure (see text)

Cascade failure

Cascade failure takes place when the failure of one component puts extra strain on other components which then successively fail as a result. A potentially very serious example of cascade failure is occasionally experienced when a surge on the electrical mains supply network can bring out a circuit breaker, thus increasing the local overload, bringing out other breakers and so on. The clear interdependency of component reliabilities in such cases can greatly reduce overall reliability.

A paper by Martin (1982) discusses cascade and other forms of failure with particular reference to mechanical systems.

Diversity

A very important technique that can be used to counter common mode and cascade failure is diversity. Frequently, redundant channels can be based on completely different physical principles. For example, a pressure vessel can be protected against accidental overpressure by the simultaneous use of a bursting disk and a pressure relief valve.

Furthermore, many navigation systems use both gyro- and magnetic compass-based elements. We will come across a number of examples of the use of diversity in the next few pages where several high integrity systems are described.

Examples of high integrity equipment

The nuclear industry took a strong lead in the development of high integrity systems and in the United Kingdom the Windscale incident in 1957 led to a further strengthening of efforts in this direction (Green and Bourne, 1966). The techniques developed for nuclear reactor systems were later applied to nuclear fuel reprocessing and then to chemical reactors.

Nuclear reactor shutdown

In a nuclear reactor the atomic nuclei of certain heavy elements are bombarded with nuclear particles called neutrons. Under this bombardment the nuclei break up, a process known as fission, and in doing so energy is released. This energy ends up as heat which can be removed from the reactor by means of a suitable coolant and used to generate electricity. The necessary neutrons are produced by the fission process itself and the rate at which fission takes place can be controlled by absorbing a greater or lesser proportion of the neutrons by inserting neutron-absorbing control rods into the reactor. The further the control rods are inserted the more neutrons they absorb, the slower the fission rate and the less the power generated.

It is most important to control the fission process in a nuclear reactor very carefully. If for some reason the process gets out of control the reactor can overheat and, in serious cases, release large quantities of radioactive material to the environment (see Chapter 17). In order to keep the likelihood of such an occurrence to a low level, an emergency shutdown system is provided. Such a system must have a very low fractional dead-time and a small probability of a spurious trip resulting from failure-to-safety.

In a real case, the shutdown system would depend very much on the detailed design of the reactor and would be far too complex to describe here. We have taken a generalized and greatly simplified example (see Figure 3.7). The number of sensing channels used, the parameter they are sensing and the number of sensors, will in practice depend on the nature of the malfunctions of the reactor for which emergency shutdown is required. In our example, we have four sensors measuring fuel temperature and three measuring the neutron intensity. These could be used to detect a rapid increase in reactor power

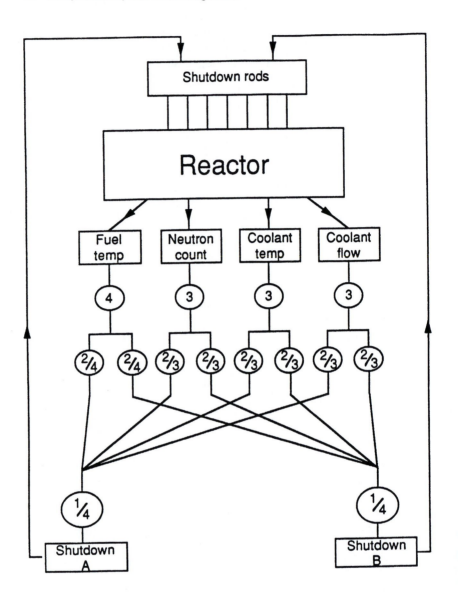

FIGURE 3.7
A simplified nuclear reactor shutdown system

following spurious withdrawal of control rods when the reactor is already operating near full power. We also have three sensors measuring coolant temperature and three measuring coolant flow in case the coolant system breaks down. Normally, the individual sensors will be independently powered to reduce common mode failure, and will be designed for failure-to-safety whenever possible. The far greater complexity of a real example would provide considerably more diversity in the detection of reactor malfunction than is present here.

Examination of the fuel temperature channel in Figure 3.7 shows that the signals indicating malfunction are fed into two paths each leading to a separate shutdown system. On each of these paths a majority vote of two out of four is needed if the signal indicating malfunction is to be passed on, thus reducing the likelihood of spurious trips following failure-to-safety of the temperature measuring devices. The other three sensing channels measuring neutron count, coolant temperature and coolant flow are similarly duplicated before being subjected to two out of three voting. The split into two paths provides parallel redundancy both on the voting systems and in the final shutdown systems. The latter are actuated on receipt of a vote signal from any one of the four sensing channels.

The reactor is shut down by the insertion either of the normal control rods or of special emergency shutdown rods. The rods are designed to be fail-safe, falling under gravity to the fully 'in' position when electrical power is removed. This is achieved by two circuit breakers in series, each powered by one of the two redundant shutdown systems, either one being sufficient to achieve total shutdown.

Chemical reactor shutdown

A number of large scale accidents took place in the chemical industry during the first half of this century, see for example Appendix 1 of Lees (1996) and the reports of the Advisory Committee on Major Hazards (Harvey, 1976, 1979). These incidents included toxic gas emissions, particularly chlorine, and several explosions involving ammonium nitrate and liquefied natural gas. These frequently involved multiple deaths. The situation got steadily more serious in later years as the chemical industry built new plants operating, for economic reasons, on a larger scale and at higher temperatures and pressures. Under such circumstances particularly close controls of the chemical reaction were essential and highly reliable means of closing the systems down to a safe state were needed.

One of the very first such shutdown systems, developed for an ethylene oxide plant, was described by Stewart (1971). Ethylene oxide is produced by a reaction between ethylene and oxygen in the vapour phase. It is essential to keep the oxygen concentration below the value at which combustion can take place and the emergency shutdown system must, when necessary, switch off the supply of oxygen to the reactor. Shutdown is initiated by direct detection of a high oxygen concentration or by detection of excessive temperature or pressure. Many secondary safety features are needed on such a reactor but these will not be discussed here.

The general layout of the system is to be seen in Figure 3.8. This is

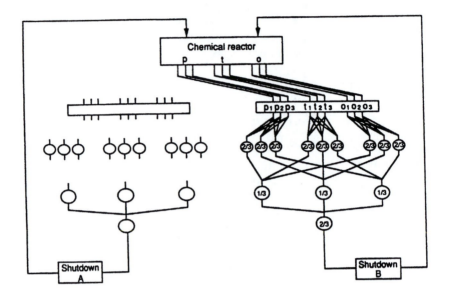

FIGURE 3.8
A simplified chemical reactor shutdown system

somewhat more complex than Figure 3.7 but the similarity of general layout should be noted. There are three parallel redundant measurements of pressure (p), temperature (t) and oxygen concentration (o) and the out-of-range signals from these are fed to two identical logic and shutdown systems A and B. System B is shown in detail in Figure 3.8.

The pressure signals p_1, p_2 and p_3 are sent to three parallel redundant voting circuits each demanding a two-out-of-three vote. The temperature signals t_1, t_2 and t_3 and the oxygen concentration signals o_1, o_2 and o_3 are similarly treated and the outputs from the voting circuits are fed to three further circuits providing a one out of three output, that is they provide a signal on receipt of either a 'p' or a 't' or an 'o' signal. These are then finally combined on a two-out-of-three vote to energize the shutdown system for the oxygen supply. The high degree of redundancy accompanied by several stages of voting provides high reliability and a strong discrimination against fail-safe spurious trips. It also makes maintenance possible while the system is in operation.

The two shutdown systems A and B each operate three shutoff valves placed in series on the oxygen input line. Three are needed to provide the degree of reliability demanded of the shutdown process. Stewart estimated the hazard rate for this system, under the particular operating conditions he specifies, to be 4.79×10^{-5} per year.

Aircraft safety

The development of high integrity systems for use in aircraft has taken place somewhat independently of that of the nuclear and chemical industries. One can trace such development as far back as the First World War when many twin winged aircraft were fitted with duplicate wing braces in order to provide some redundancy in that respect. The airworthiness requirements in the 1940s and 1950s were based on specification of engineering detail for each functional subsystem (Lloyd and Tye, 1982). The importance of redundancy and diversity were well understood and were used in the specifications to provide adequate reliability. The rapid development of aircraft design in later years made this approach increasingly difficult to use. This was particularly so in view of the considerable interdependence which developed between functional subsystems as a result of the introduction of auto-stabilization, automatic landing and other such devices.

Aircraft designers utilize many of the techniques, both qualitative and quantitative, that are used in the design of other high reliability equipment. Such techniques are employed nowadays to demonstrate predicted compliance with airworthiness requirements which are usually expressed in terms of required performance, reliability and availability. The result, in terms of increasing aircraft safety, has been impressive.

Figure 3.9 shows somewhat schematically a possible configuration of the elevators, the ailerons and the spoilers on a modern civil passenger aircraft as described, for example, by Lloyd and Tye (1982). All these control devices are subdivided in order to provide parallel redundancy, the elevators and ailerons into two, the spoilers into three. The spoilers are hydraulically controlled, while the other devices are electrically controlled although using local hydraulic power to apply the control forces. Both electrical and hydraulic power are provided by the four aircraft engines. The notation used in Figure 3.9 indicates E for electrical power and H for hydraulic power, followed by a number indicating which engine provides the power. Thus, the outer aileron on the port wing (right hand side of Figure 3.9) is powered electrically from engines one and three, the inner by engines two and four. Again we see obvious parallel redundancy, any one engine being able to provide sufficient power on its own. The hydraulic power contains similar redundancy. Reconfiguration of the electrical supplies is possible so that, for example, if E1 and E3 both fail, power can be provided by E2 and E4. In the ultimate extremely unlikely situation where all the engines fail, many aircraft can generate a limited amount of power from a small ram jet turbine which can be lowered into the aircraft slipstream. Such a standby device gives diversity in power provision. A further level of diversity is present in the sense that limited control of aircraft manoeuvrability can be affected in the

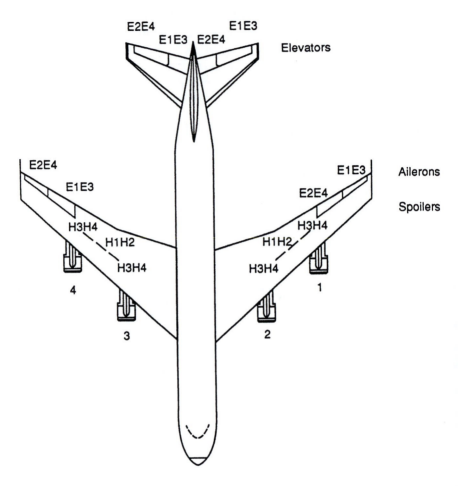

FIGURE 3.9
The flight controls of a civil passenger aircraft

event of total loss of electrical power by using the hydraulically powered tail trim and spoilers.

An important potential source of common mode failure in a wide range of high integrity equipment is damage to power or control lines. This is a particular problem in aircraft due to their attenuated shape and the lack of space normally available. The damage which might, for example, simultaneously affect power lines to parallel redundant units could be caused by fire or explosion, but might also be due to less dramatic events like the ingress of water, loss of hydraulic fluid or chemical corrosion. A particular hazard to power and control lines in aircraft is provided by the possible tangential shedding of turbine components from the engines.

Only relatively large fragments would be able to penetrate the body

FIGURE 3.10
Trajectories of engine fragments

of the aircraft, and by examining the many possible trajectories of such fragments (Figure 3.10) it is possible to find routes such as those marked XX which minimize the chance of simultaneous damage.

Conclusions

In this chapter we have seen how high reliability is attained and we have examined some simple examples of high reliability equipment. We now proceed to study how such equipment is designed and how identification and analysis of hazards (hazard analysis) can lead to design improvements, thus giving greater reliability.

Further reading

Green, A. E. (ed.) (1982) *High Risk Safety Technology*, John Wiley, Chichester.

Klaassen, K. B. and van Peppen, J. C. L. (1989) *System Reliability, Concepts and Applications*, Edward Arnold, London.

Lloyd, E. and Tye, W. (1982), *Systematic Safety*, Civil Aviation Authority.

References

Andrews, J. D., and Moss, T. R. (1993), *Reliability and Risk Assessment*, Longman Scientific and Technical.

Edwards, G. T. and Watson, I. A. (1979) *A Study of Common Mode Failures* (Safety and Reliability Directorate Report SRD R14, July 1979), United Kingdom Atomic Energy Authority.

Green, A. E. and Bourne, A. J. (1966) United Kingdom Atomic Energy

Authority Report AHSB(S) R117.

Green, A. E. and Bourne, A. J. (1972) *Reliability Technology*, Wiley Interscience, Chichester.

Harvey, B. H. (1976) *First Report of the Advisory Committee on Major Hazards*, HMSO, London.

Harvey, B. H. (1979) *Second report of the Advisory Committee on Major Hazards*, HMSO, London.

Lees, F. P. (1996) *Loss Prevention in the Process Industries* (2nd ed.), Butterworth-Heinemann, Oxford.

Lloyd, E. and Tye, W. (1982) *Systematic Safety*, Civil Aviation Authority.

Martin, P. (1982) Consequential Failures in Mechanical System, *Rel. Eng.*, **3**, 23-45.

Stewart, R. M. (1971) *High Integrity Protective Systems*, Inst. Chem. Engs. Symposium Series No. 34.

Chapter 4

Failure to danger

Introduction

In Chapter 3 we considered several high integrity systems in which redundancy and diversity were used to reduce failure to danger to extremely low values. For such a system to be effective, a good basic design is necessary. We consider first some of the factors to be taken into account in producing such a good basic design. We then go on to examine techniques used to identify failure modes and to predict failure probabilities in high integrity systems.

The design procedure

Production of the final design of a system will normally be an iterative process in which:

1. A basic sound design intended to meet the design specification criteria is produced.
2. The proposed design is then analysed to see what failure modes are present.
3. Probabilities of failure by these modes are calculated.
4. Where safety criteria are involved, the hazards resulting from the failures predicted. Improvements are then incorporated where appropriate and the design is reanalysed.

Such a procedure, known as hazard analysis, leads to design improvements thus giving greater reliability. It can also be used to mitigate the more serious effects to be experienced on failure.

Hazard analysis will necessarily involve an understanding of the physical and sometimes the chemical properties of the materials involved. We will spend some time discussing these properties, but first it is important to distinguish between intrinsic and extrinsic safety.

Intrinsic and extrinsic safety

Kletz (1978) has given good examples of intrinsic and extrinsic safety in the design of a house. One of the features causing most accidents in a house is the stairs. The extrinsic design solution is to specify a good firm hand rail and ensure that the stair carpet is in good order and securely fastened. Regular maintenance will obviously be needed in order to ensure reliable functioning of these safety-related features, but even then some accidental falls will be inevitable. The intrinsic solution is to design a single storey dwelling at ground level. Note that the improvement introduced by the intrinsically safe solution is only in relation to one hazard. For example, both dwellings will still have cooking and heating facilities based on electricity or gas with their associated hazards. The intrinsic solution is the better one, and is used wherever possible. The extrinsic solution makes use of 'add-on' features designed either to make failures less likely or to lessen their effects.

Many examples of both the intrinsic and the extrinsic approach to safety are to be found in the field of chemical engineering. For example, many of the products of the chemical industry are relatively inert and chemically stable, but they may be produced in a multi-step process involving highly reactive or toxic intermediates. An alternative intrinsically safe process avoiding such intermediates can sometimes be found, otherwise it is particularly important to limit inventories in such circumstances. The hazard can also sometimes be reduced by lowering temperatures and pressures, but it is not possible to generalize because inventories will then frequently have to be increased in order to maintain output at the resultant reduced reaction rates. The hazardous properties of some reactive compounds can also be attenuated by dissolving them in a harmless liquid.

Mechanical systems and design

Sound mechanical design is of fundamental importance in almost every situation where reliability is at a premium. Such design must be based on an understanding of how mechanical failure can come about and how it can be avoided.

Mechanical stress

In Figure 4.1 we see the typical behaviour of a steel sample as a mechanical load is applied. The figure displays the stress (stretching force per unit area) on the vertical axis against strain (fractional increase

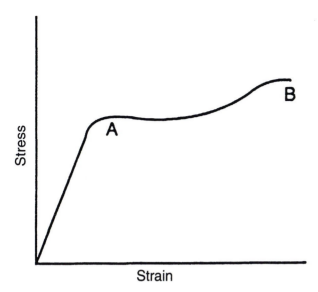

FIGURE 4.1
The stress versus strain curve for a steel sample. Point A is the yield point, point B the rupture point

in length). Following the initial linear region where stress is proportional to strain, the point A, known as the yield point, is reached. Beyond A the material will extend without further increase in stress. If more stress is applied, however, point B is eventually reached at which the sample ruptures. The stress value at B is known as the ultimate tensile stress. The yield point stress and the ultimate tensile stress are both important parameters in the design of mechanical systems such as support structures and pressure vessels. Clearly, the latter parameter cannot be exceeded and it is normal in addition to design mechanical systems such that the maximum stresses in the bulk of the material are significantly below the yield point stress. This stress value may well be exceeded over small regions however, due to local stress concentrations.

Fatigue failure

The most common mode of failure for pressure vessels is by cracking. The cracks can be due to welding flaws, material porosity or inclusions. Cracks can also be caused by metal fatigue. This phenomenon occurs when material is subjected to a cyclic variation in stress. Failure may eventually take place even though the peak stress is well below the ultimate tensile stress.

The fatigue performance of steel is illustrated in Figure 4.2 (curve (a)). The plateau represents a stress cycle amplitude, known as the fatigue limit, below which the material can be cycled indefinitely without failure. Where fatigue effects may be present the design stress is normally kept below the fatigue limit to avoid fatigue failure.

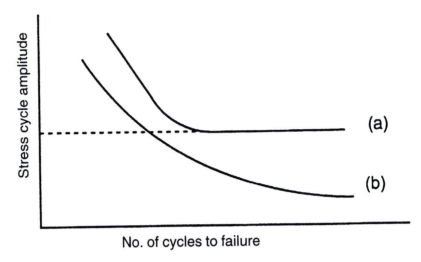

FIGURE 4.2
The fatigue performance of (a) steel and (b) aluminium alloy

For an aluminium alloy (curve (b)) there is no plateau and the material will eventually fail, however small the stress cycle amplitude. Such stress cycling could be due to the filling and emptying of a pressure vessel, to thermal cycling, or, as in the case of the Comet aircraft tragedies in the 1950s, to atmospheric pressure cycling as the aircraft gained and lost altitude (see, for example, Pugsley, 1966).

Crack propagation

Before a crack can cause failure it must reach a certain critical size, this size is dependent on the elastic properties of the material and the magnitude of the stress. The higher the stress, the smaller the critical crack size. The crack can grow to the critical size as a result of fatigue effects or under the influence of creep. Creep is a phenomenon producing permanent distortion under the action of a steady load at elevated temperature. Such temperature conditions are present in many engineering structures where failure could have serious effects.

The sequence of events once the critical crack length is reached depends on the nature of the load and on the dimensions and elastic properties of the material. There may be fast fracture which can attain explosive violence or a much slower incremental tearing. The Comet failures provided a good example of fast fracture. Incremental tearing provided a serious hazard in a number of freight ships of welded construction used during World War II.

A number of chemical and physical processes of direct relevance in modern high-risk industries can have a profound effect on the mechanical properties of materials. The presence of corrosive chemicals can greatly enhance crack formation and growth, while both hydrogen absorption and nuclear radiation can cause embrittlement. An understanding of fracture mechanics is very important in the provision of reliable mechanical design. It is difficult to generalize as to how this knowledge is used but clearly any design must provide an acceptable lifetime against fatigue and must be such that regular inspection will reveal crack propagation before the cracks can reach critical length. We have only been able to discuss fracture mechanics in barest outline. The reader is referred to other texts such as Chapter 12 of Lees (1996) or Chapter 4 of Thomson (1987) for more detail.

Chemical systems and design

In dealing with chemicals, either on a laboratory or a commercial scale, hazards of fire, explosion and of toxicity are frequently encountered and these must be taken into careful consideration at the design stage. This section describes how the associated properties of the chemicals are defined.

Fire hazards

A gas or vapour will only burn in air within a limited range of concentration. A mixture that is too weak or too rich will not ignite.

We can define corresponding lower and upper flammability limits in air, quoted as percentages by volume. For hydrogen, for example, the lower limit is 4% by volume and the upper limit is 75%. For many flammable vapours the upper limit is much lower but for acetylene and ethylene oxide it is 100% (that is, no air need be present for fire to propagate). In general, the wider the upper and lower limits are spaced the greater the fire hazard. The limits are affected by a number of factors, including temperature, pressure and the presence of inert gases (see, for example, Chapter 16 of Lees, 1996). Similar limits, usually more widely spaced, can be defined for combustion in pure oxygen.

The flash point is another important parameter. This is the temperature at which the vapour pressure at the surface of the liquid is high enough for the lower flammable limit to be reached. At lower temperatures the air vapour mixture is, in principle, too weak to ignite. The auto-ignition temperature is the temperature at which bulk combustion occurs. Typically, it is in the range 200°C to 500°C for various flammable organic chemicals. If an ignition source is applied, a minimum ignition energy is found to be required. At optimum concentration this is typically 0.25 mJ but is as low as 0.019 mJ for acetylene and hydrogen, which means that particular care must be taken in handling these gases. The assumption that all ignition sources can be eliminated is a dangerous one in any circumstances. The safe approach is to control flammable materials such that mixtures remain outside the flammable limits.

Explosion

An explosion is a rapid release of energy either in a confined or an unconfined space. One source of such energy is a flammable gas or vapour. The release may be in the relatively slow deflagration process in which the speed of advance of the flame is typically 1 metre per second. Alternatively a detonation may take place. In this case a shock wave is built up which can travel at several thousand metres per second. Detonation is usually more destructive than deflagration, producing significantly higher over-pressures. Detonation limits defined in a similar way to the flammable ones are frequently quoted for flammable gases and vapours. These limits are narrower than the flammability limits. Protection against explosion can take the form of pressure systems designed to withstand the associated over-pressure, bursting panels or disks which allow safe depressurization, explosion suppression systems which detect the beginning of the pressure rise and rapidly apply a suitable suppressant and blast walls. A well known recent example of an accident involving the explosion of flammable vapour was that at Flixborough in 1974 in which 29 people were killed (see Chapter 17).

Toxicity

Many substances are toxic, i.e. they are harmful to health if they enter the body, a process that may take place by inhalation (breathing), ingestion (swallowing) or by absorption through the skin or other tissues. In order to provide against airborne toxic substances in the workplace, occupational exposure limits are defined which provide maximum permissible concentrations of such substances. In the United Kingdom the Health and Safety Executive (HSE) has set maximum exposure limits (MELs) and less strict occupation exposure standards (OESs) for long-term (eight hours) and short-term (usually fifteen minutes) exposure. Long-term limits are designed to guard against slow accumulation in the body or the development of chronic disease. Short-term limits protect against irritation of lungs, eyes or skin, or other acute effects including serious injury or death. The HSE limits are published annually in their Guidance Note EH40 (HSE, 1995).

Limits are quoted for dusts and fumes as well as vapours. Maximum acceptable concentrations are expressed as parts per million in air or as milligrams per cubic metre of air. For example, the long-term MEL for carbon disulphide is 10 ppm or 30 mg m^{-3}. Fibrous dusts such as asbestos are given limits quoted in numbers of fibres per millilitre of air. Where concentrations vary, time weighted averages need to be determined.

A substance is assigned an OES if it is possible to identify a concentration at which there is no indication that inhalation day after day is likely to be injurious. Furthermore, where exposure to higher concentrations might occur, serious short- or long-term effects must not be expected over the timescale required to identify and remedy the cause of excessive exposure, assuming action is taken as soon as is reasonably practicable.

Where a substance does not satisfy these criteria or where a higher concentration is to be adopted for other reasons, then an MEL is allocated. Exposure must be below this level and in fact will only be considered as acceptable if reduced as far as is reasonably practicable below the MEL in each particular case. In setting an MEL, socio-economic factors are taken into account. The UK Health and Safety Commission is currently considering using cost-benefit analysis as part of this process (HSC, 1996).

The best method of limiting exposure is by total enclosure of the source. Where this is not possible, a partial enclosure with carefully controlled air movement, as for example in a fume cupboard, can be used. Personal protective equipment such as respirators or breathing apparatus, or the limitation of exposure by control of time on the job provide less desirable solutions. In some situations good general ventilation may prove adequate.

In the United States, the Conference of Governmental Industrial Hygienists (ACGIH, 1981) issues a very extensive list of exposure limits, known as threshold limit values (TLVs), with similar definitions to those discussed above. In addition, various United States Governmental Agencies provide limits covering chemicals which can cause environmental damage. A good general description of TLVs and their use has been given by Doull (1994).

If a serious fault develops in chemical plant, exposures greatly in excess of the occupational exposure limits may be encountered. Methods used to predict the effects of large exposures are discussed more fully in a later chapter. They will obviously depend on the chemical or chemicals involved and also on the route of intake to the body. The LD_{50} is the dose that is predicted to prove lethal to 50% of those exposed while the ED_{50} is that which will be effective in producing a particular condition or symptom at the same 50% level. In studying environmental effects the LC_{50} is employed, this being the lethal concentration in air or water to 50% of the exposed population in a given exposure time which may be a few hours or a number of days.

Other properties

Many other properties must be taken into account, where relevant, at the design stage. For example, components in electrical equipment must have adequate power ratings and sufficient cooling must be provided at the right places to ensure satisfactory operating temperatures. Electrical insulation must be sufficient for the voltages involved. For equipment involving ionizing radiation, x-ray machines, accelerators used for medical treatment and nuclear reactors, for example, adequate shielding must be provided against the radiation hazards. Ionizing radiation is discussed in more detail in Chapter 10.

Identification of hazards

In previous sections we have considered a number of mechanical, physical and chemical properties which can present a hazard under fault or failure conditions. It is not always possible to be sure that all situations have been taken into account, however, and various techniques are available to help in this respect.

Many hazards, particularly in relatively uncomplicated situations, can be immediately apparent. In other cases they are not so obvious, but as long as they are not too serious we can proceed by trial and error, making corrections as we go along – 'every dog is allowed one bite' – Kletz (undated). In situations where major hazards are involved such a procedure is simply not acceptable and a more systematic approach must be taken.

The simplest systematic approach is to use some form of check list. This practice is quite common and works well in situations where the configuration is similar to previous ones and the hazards are also similar. For a completely new design the checklist will be of limited applicability.

Hazop

The hazard and operability study (Hazop) was developed in the United Kingdom at ICI in the 1960s from the 'critical examination' technique then in use, (Houston, 1971). Hazop provides a systematic way of identifying hazards using a number of guide works as an aid. Just how the words are interpreted depends on the circumstances, but for example 'None of' could lead to a consideration of the possibility of no liquid flow in one case or no electrical current or no pressure in others. Other guide words and some applications are:

'More of'	– liquid flow too high, temperature, pressure or electrical current too high.
'Less of'	– liquid flow too low, temperature, pressure or electrical current too low.
'Part of'	– chemical component missing, composition wrong.
'More than'	– impurities present, extra phase present (gas in liquid, for example).

In each case the guide word is used to concentrate attention on to one particular fault (no liquid flow, for example). Possible causes of lack of flow are then examined and the effects of it are enumerated. The complete set of guide words is applied in this way to each component or process in turn. The design can then be modified to avoid the associated hazards and consequent operational problems.

The technique is commonly used in the chemical industry and is particularly effective if applied by a mixed team providing expertise in design, instrumentation, commissioning and operation. Andrews and Moss (1993), Kletz (undated) and Lees (1996) provide good examples of the use of Hazop procedures.

Zonal analysis

Zonal analysis is a method used to examine possible cascade and common mode failures in aircraft. In this case the aircraft is sub-divided into zones and for each zone all actuators and other items of equipment within the zone are itemized. The mutual interactions within the zone are then examined as are the interactions with similar devices outside the zone. The interactions may be anything from electrical interference to leakage of hydraulic fluids or water, or undesirable mechanical interactions. Both normal and fault conditions are considered. As in the case of Hazop, zonal analysis provides a systematic framework for the investigation of a particular type of failure.

Hazard analysis

Having identified possible hazards and how they might materialize, it is necessary to provide a design that will reduce them to a predetermined level. In cases where the hazards are well understood and similar apparatus has been used before, codes of practice can be very useful. Many examples of such codes can be quoted going back well over a hundred years. For instance, many accidents occurred in the early days of railways due to boiler explosions caused by faulty design or the lack of an adequate pressure relief valve to prevent accidental over-pressurization. Accidents were greatly reduced in frequency once adequate codes of practice were introduced for the design and operation of such pressure vessels. Again, codes of practice were extensively used in the early days of aircraft construction. Such codes may be issued by the safety authorities or by learned institutions or other professional groups. With the very rapid later development of aircraft of increasing complexity, other methods were found necessary to validate design.

The other methods involve the prediction for a given equipment of the likelihood of hazards materializing by systematically examining all the individual faults or combinations of faults which can lead to such hazards. Many of the techniques used in estimating this likelihood also contribute to the identification of possible hazards, so the distinction between the two processes is not as clear cut as it might seem.

The techniques employ two general approaches. In one, known as the 'top down' approach, we start from a particular hazardous outcome (a loss of aircraft control or a toxic release, for example) and work backwards through the failures or combinations of failures that could lead to that final event. In other words we move from effect to cause. The alternative approach is the 'bottom up' approach in which we start from a specific failure and follow this through to trace all possible hazardous outcomes, thus moving from cause to effect.

Failure modes and effects analysis

An important example of the 'bottom up' approach is the failure modes and effects analysis (FMEA). This is of particular use in examining the performance of relatively simple components, for determining which types of failure are to danger and which are to safety, and finally for calculating overall failure rates to the two states for the complete component. The example we are going to discuss (an FMEA study of a pressure switch) has been used for many years and was introduced by Hensley (1970). The mechanism is illustrated in Figure 4.3 and is designed to give an indication when gas pressure is lost.

The pressure p is applied as in Figure 4.3(a) to a bellows which expands pressing up the pivoted bar which closes the electrical contacts. If pressure is lost (Figure 4.3(b)) the bellows contracts and the action of the spring on the pivoted bar breaks the electrical contact, a change that is used to give warning of pressure loss.

Rupture of the bellows gives failure to safety as it would lead to loss of pressure in the bellows and an indication at the electrical contacts. On the other hand, loss of the pivot (Figure 4.3(c)) or fracture of the spring (Figure 4.3(d)) give failure to danger. In both cases the electrical contacts are held closed even if pressure is lost, so no loss of pressure signal would be provided. The situation with the electrical switch is less clear. The fact that the contacts are opened when pressure is lost makes the switch fail to safety if the electrical wires or connectors are broken. On the other hand, failure of the spring in the switch could leave the contact open or closed depending on the exact form of failure. We assume in the circumstances that failure to safety will occur in 75% of occasions, failure to danger in the other 25%.

Using total failure probabilities (reliabilities) for a one-hour period quoted by Green and Bourne (1973) we now have the situation of Table 4.1.

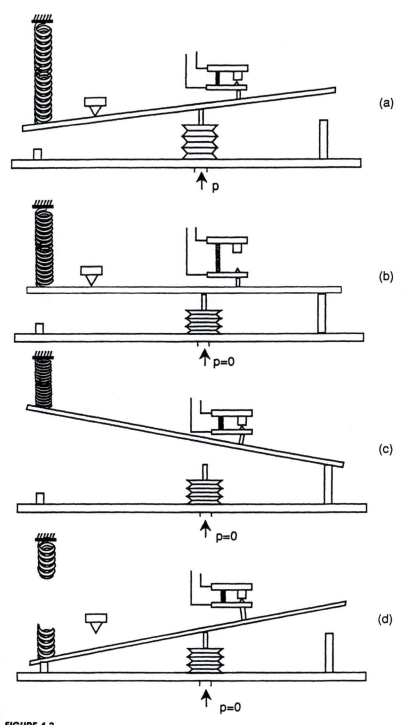

FIGURE 4.3
Schematic diagram of a pressure switch. (a) Pressure applied; (b) pressure removed; (c)
pivot displaced; (d) return spring fractured

Table 4.1 Pressure switch failure

Item	Failure probability	Failure to danger probability	Failure to safety probability
Bellows	5.0×10^{-6}		5.0×10^{-6}
Pivot	1.0×10^{-6}	1.0×10^{-6}	
Spring	0.2×10^{-6}	0.2×10^{-6}	
Switch	2.0×10^{-6}	0.5×10^{-6}	1.5×10^{-6}
Total	8.2×10^{-6}	1.7×10^{-6}	6.5×10^{-6}

Thus we have a failure to danger probability of 1.7 210^{-6} and a failure to safety probability of 6.5 210^{-6}. In general, the failure to danger probability will be of relevance to hazard analysis and the failure to safety probability is needed in the calculation of equipment reliability and the effect of spurious trips. Note that failure probabilities have been added. In the pressure switch we have a typical series system – any one component failing will mean failure of the device. In addition, the component reliabilities are assumed independent of each other. Thus we have from Lusser's law, using the notation of Chapter 2:

$$R = R_1 \times R_2 \times ... \times R_n$$

or

$$1 - F = (1 - F_1) \times (1 - F_2) \times ... \times (1 - F_n)$$

For the case where there are only two components:

$$1 - F = (1 - F_1) \times (1 - F_2)$$
$$= 1 - (F_1 + F_2) + F_1 F_2$$

But in the present case F_1 and F_2 are very small compared with 1 (and usually are in most cases we are likely to study), so the term $F_1 F_2$ will be negligible in comparison to $(F_1 + F_2)$. Therefore we can omit it such that:

$$1 - F = 1 - (F_1 + F_2)$$

i.e. $F = F_1 + F_2$

Similarly in the more general case:

$$F = F_1 + F_2 + ... + F_n$$

Thus the overall failure probability is the sum of the component ones. This is how we arrived at the total failure to danger and failure to safety probabilities in the example of the pressure switch.

Event tree analysis

A second 'bottom up' technique, again starting from a particular component failure and following through to trace possible resulting hazards, is event tree analysis. This is used most frequently for rather more complex systems than is FMEA. Our example involves a study of what might happen on the failure of the off-site electricity that powers the controls of a nuclear reactor. The initiating event, as it is called, is thus 'loss of off-site electricity' (Figure 4.4).

The first safety feature that should come into operation on loss of electrical power is the shutdown of the reactor using a system similar to that described in Chapter 3. The shutdown rods are designed to operate automatically on loss of power.

An exercise similar to the FMEA just described would provide a reliability R_{RT} for the reactor trip system and a probability $P_1 = 1 - R_{RT}$ that the trip system would fail (Figure 4.4). In the latter case the resulting effect on the reactor would be sufficiently severe that later safety features would be irrelevant.

Following a successful reactor trip with probability R_{RT}, the reactor design provides for activation of emergency electrical supplies. If they fail to come on (probability $1 - R_{ES}$), the emergency core cooling cannot operate and we assume that the primary containment surrounding the reactor will be breached, although the secondary containment may not be, with probability $1 - R_{SC}$. We can calculate the probability P_2 of a

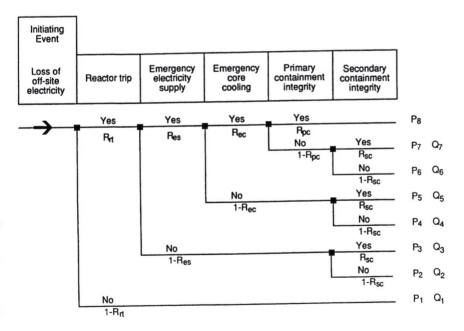

FIGURE 4.4
Event tree analysis – nuclear reactor power loss

breach using Lusser's law, as the probabilities involved are independent. Thus in Figure 4.4

$$P_2 = R_{RT} \times (1 - R_{ES}) \times (1 - R_{SC})$$

Similarly, the probability that the secondary containment will not be breached will be

$$P_3 = R_{RT} \times (1 - R_{ES}) \times R_{SC}$$

If the emergency electrical supply is successfully activated with probability R_{ES}, power is available for the emergency core cooling system. This is needed as a significant amount of heat continues to be generated in the reactor core even after shutdown, due to radioactive decay processes. If the cooling system fails to operate (probability $1 - R_{EC}$) we assume that the primary containment will be breached but that the secondary containment may or may not be effective, leading to overall probabilities of P_5 and P_4 respectively.

With emergency core cooling in operation the primary containment has a probability R_{PC} of not failing and in that case the secondary containment is not needed. If it does fail we have probabilities P_7 and P_6 respectively that the secondary containment will or will not hold. P_4 P_5 P_6 P_7 and P_8 are calculated in a similar way to P_2 and P_3.

Note that the event tree contains an implicit time factor in that the events follow each other from left to right. Some of the probabilities involved in the calculations can be evaluated relatively easily (R_{RT} and R_{ES} for example) while others like R_{PC} raise considerable difficulties. Indeed much of the criticism of hazard analysis is centred on such difficulties as we will see in later chapters.

Our example is a grossly over-simplified one. For example, we have assumed the same values of R_{SC} in the three different situations where secondary containment is involved. Again many more logical steps could be included in the event tree to take into account, for example, the two-stage process of bringing on the emergency electrical supplies, to account for reintroduction of off-site supplies following failure or to allow for the effects of delay in the start-up of the emergency core cooling. At a more fundamental level our assumptions about when containment might or might not be breached are somewhat simplistic. It is hoped, however, that the example provides an explanation of how the event tree is assembled and used. Further examples are found in the scientific literature.

Fault tree analysis (HAZAN)

Fault tree analysis (or HAZAN) provides a powerful and commonly used example of a 'top down' procedure. The example we give is based on a real study (Hensley, 1981) into the handling of containers holding highly radioactive spent nuclear fuel located under water in a fuel

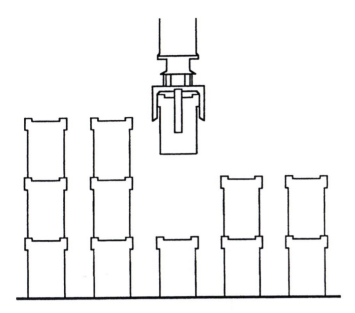

FIGURE 4.5
Container handling system with three-high stacks

storage pond. A similar analysis can be made for an ordinary container handling system as described here although a collision between containers would be more likely in the nuclear case unless great care is taken, as fuel container movements within the fuel storage pond can only be observed indirectly using closed-circuit TV.

In the fault tree analysis we start from a hazardous outcome, known as the top event. In the present analysis the top event is the overturning of a container during the movement by means of the lifting mechanism of another container.

The situation is illustrated in Figure 4.5 where the lifting mechanism is shown at its highest point. In other words, it is able to stack containers three high but not to lift containers over a three-high stack. The lifting mechanism has three safety interlocks. Two working in parallel detect excessive lateral forces and remove power for lateral movement if this is detected. The third ensures that retraction of a container from its position in the stack is complete before lateral movement is allowed. Thus it cannot move sideways until clear of the container below.

The fault tree is shown in Figure 4.6 which is a logic diagram indicating the sequence of events leading to the top event, the overturn of the container. Note the use of domed 'AND' gates and pointed 'OR' gates. Thus for the container to overturn we require a collision AND the availability of sufficient force. The latter will only be present if both lateral interlocks, that is interlock 1 and interlock 2 fail.

The conditions for a collision are the movement of a container unretracted OR a collision with a triple stack. In the latter case the stack

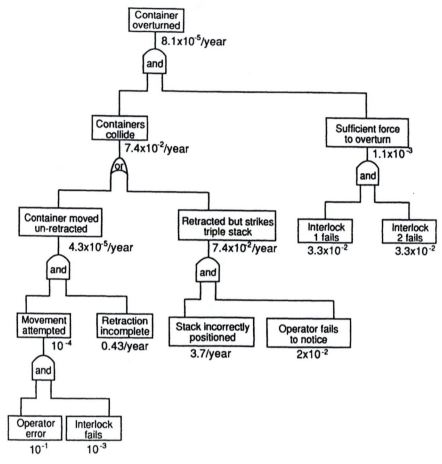

FIGURE 4.6
Fault tree for overturn of a container

must have been incorrectly placed on the position AND the operator must fail to notice the error. For the container to be moved while unretracted, the retraction must be incomplete AND movement must be attempted, while for movement to be attempted in these circumstances, there must be an operator error AND a failure of the relevant interlock. The logic just described can easily be followed in Figure 4.6.

Having established the logic diagram we can now, given the appropriate component probabilities, calculate the probability of the top event. The rules for combination of probabilities P_A and P_B are as discussed, for example, in Chapter 7 of Lees (1996):

$$P(A \text{ and } B) = P_A \times P_B$$
$$P(A \text{ or } B) = P_A + P_B - P_A P_B$$
$$= P_A + P_B \text{ if both } P_A \text{ and } P_B \text{ are much less than unity}$$

The component probabilities used in Figure 4.6 are taken from Hensley (1981). They have been combined at AND and OR gates using the rules quoted above. In some instances, event rates (events per year in this case) have been quoted instead of probabilities. This is only valid under certain circumstances and is discussed in the recommended texts (see Further Reading).

Figure 4.6 indicates that the probability over a one-year period of the overturning of a container is 8.1×10^{-5}. Note that the collision probability is dominated by the triple stack collision because of the presence of the interlock on the retraction process. Without this the probability for a one-year period of a container being moved unretracted would be 4.3×10^{-2}, and the probability of the containers colliding would be

$$= (4.3 + 7.4) \times 10^{-2}$$

or:

$$= 1.17 \times 10^{-1}$$

The fault tree, unlike the event tree, represents a purely logical system containing no time element, either implicit or explicit. Our example is a very simple one. In practice, fault trees can be very much more complex and can contain other types of logic gate. The kinetic tree approach, for example, introduces time variation. Hazard analysis techniques are discussed in more detail in Andrews and Moss (1993) and Lees (1996).

Conclusions

Hazard analysis is a complicated procedure which employs a wide range of techniques. Some of these have been discussed in this chapter by way of illustration. The discussion has been very brief, however, and should not be used as a basis for the employment of the techniques.

We have seen how known component reliabilities are used to predict an overall reliability which may frequently be so small that it is impossible to check the accuracy of prediction by direct observation. In these circumstances total reliance has to be placed on the accuracy and completeness of the logic diagrams and on the relevance and accuracy of the component reliability data used. In socio-technical systems it is also dependent on the reliability of the human component. This is discussed in the next chapter.

Further reading

Green, A. E. (1983) *Safety Systems Reliability*, John Wiley, Chichester.
Green, A. E. (ed.) (1982) *High Risk Safety Technology*, John Wiley, Chichester.

Thomson, J. R. (1987) *Engineering Safety Assessment*, Longman.

References

American Conference of Government Industrial Hygienists (1981) *Documentation of the TLVs*, ACGIH, Cincinnati.

Andrews, J. D. and Moss, T. R. (1993) *Reliability and Risk Assessment*, Longman Scientific and Technical, Marlow.

Doull, J. (1994) Threshold Limit Values, how they are established and their role in the workplace, *Inhalation Toxicology*, 6(Suppl.), 289.

Green, A. E. and Bourne, A. J. (1973) *Reliability Technology*, Wiley Interscience, Chichester.

Hensley, G. (1970) *Nuc. Eng. Design*, **13**, 222.

Hensley, G. (1981) *Proc. Symp. on Industrial Risk Assessment, Safety and Reliability Society*, Southport, UK, September 1981, p. 1.

Houston, D. E. L. (1971) *Inst. Chem. Eng. Symposium Series 34*.

HSC (1996) Proposals for amendments to the Control of Substances Hazardous to Health Regulations 1994 and the Approved Code of Practice: Control of Substances Hazardous to Health (General ACOP).

HSE (1995) *Occupational Exposure Limits*, Guidance Note EH40/95.

Kletz, T. A. (1978) What you don't have, can't leak, *Chemistry in Industry*, 6, May.

Kletz, T. A. (Undated) *Hazop and Hazan*, Inst. of Chem. Engs.

Lees, F. P. (1996) *Loss Prevention in the Process Industries*, 2nd ed., Butterworth-Heinemann, Oxford.

Pugsley, A. G. (1966) *The Safety of Structures*, Edward Arnold, London.

Thomson, J. R. (1987) *Engineering Safety Assessment, An Introduction*, Longman Scientific and Technical, Marlow.

Chapter 5

Human reliability: quantitative and qualitative assessment

Introduction

Human reliability should be an important consideration in all stages of systems design and in its later implementation and management. It can be enhanced by appropriate selection, training and continuing development of all personnel (see Chapter 7). Additional enhancement may be obtained from heightened awareness, in appropriate staff, of the design features of plant and equipment which may lead either to operational or maintenance errors. For example, instrumentation and control systems should be designed so that operators, both as individuals and when working as a team, have adequate information and time to make decisions. Control room layout should accommodate the operators and should be arranged to minimize the possibility of errors (Raafat, 1983; Hollnagel, 1993).

However, even when selection and training are efficiently carried out and appropriate design features are incorporated, people are not always reliable. They make mistakes and in some cases their errors will lead to systems failure and accidents. Early reports produced by the United States Air Force indicate that human error was responsible for a large proportion of aircraft accidents, 234 out of 313, during 1961 (Willis, 1962) and empirical and analytical studies have shown that human error contributes significantly to the accident risk in nuclear power plant operation (INPO, 1985; Barnes, 1990). The Health and Safety Executive (HSE) claim that human error contributes to as many as 90% of workplace accidents in the United Kingdom (HSE, 1989). They also indicate that as many as 70% of such accidents may be preventable. This attribution of accident causation to errors and unsafe behaviours rather than unsafe conditions tends to oversimplify human reliability. It also underestimates the importance of task and environmental variables in creating error-provoking situations.

Human reliability specialists have attempted to incorporate human behaviour into a suitable framework for the analysis of systems reliability. In doing so their primary focus has been the quantification of human error. In its common usage, the term human error is ambiguous and it generally carries a negative attribution. It is used to cover many different situations and events, including management decision errors, design and maintenance errors, but most particularly operator errors (Watson and Oakes, 1988). Major incidents such as Three Mile Island (1979), Bhopal (1984), Chernobyl (1986), and the Zeebrugge Ferry disaster (1987) have not only emphasized the importance of each of these types of error, but also their interaction in the accident process (see Chapter 17). For example, in 1979, the Kemeny Report described the interplay of a large number of managerial, organizational and regulatory root causes in relation to the Three Mile Island accident (see Rubinstein and Mason, 1979). Perrow (1984) and Kletz (1985) have also both provided numerous case studies which illustrate the ways in which human error at various levels in an organization can give rise to major 'system' disasters.

Somewhat later, Reason (1989a, b) went on to argue for a more integrated approach to accidents and errors made necessary by these disasters in complex but seemingly well-defined systems. He suggested that the root causes in all cases appeared to be 'latent' in the organizations and in the design and management of the systems, long before a recognizable accident sequence could be identified. These 'latent' problems do not appear to belong exclusively to any one domain (hardware, software or people), rather they emerge from the complex but as yet little understood interaction between these different aspects. This view has been extended by one of the present authors in a review of 'stress, cognition and control room operations' (Cox et al., 1990).

Two points result from the acceptance of the model implicit in these arguments. First, the accident process can be usefully described, in the terms of general systems theory, as an interaction between factors at several different levels of analysis, individual, social (team), organizational and technical (Cox and Cox, 1996). This interaction can result in 'latent' accidents (accidents waiting to happen) which in turn reflect or are triggered by a range of circumstances including operator error. Second, reliability specialists should consider the interplay between these different factors and their context on one hand, and the triggering events on the other.

While an argument is often made for fully automated systems (no potential for, at least, operator error), these have not always been successful in safety terms or acceptable to the public or client groups (for example, in transport systems). Furthermore, even automated systems need constant monitoring and maintenance and have their own reliability problems (see Chapter 8). It is interesting that there has also been a disproportionate increase in incidents leading to injury during maintenance tasks (HSE, 1985).

In order to extend our understanding of systems reliability, we need to examine what is involved in *human reliability* and *error*. In particular we need to:

1. Classify and understand common error types as trigger events.
2. Explore human reliability in a systems context.
3. Quantify human reliability to allow some assessment of its role in determining a system's overall reliability; i.e. probability of a trigger event in the 'liveware'.
4. Make the knowledge and experience of experts in human reliability analysis available to those who need it and are thus able to enhance human reliability, e.g. managers or decision makers and designers.

This chapter will outline some of the current understanding of human reliability and error, and make reference to points 1–4 above. It will discuss the concept of human error probability and refer the reader to more detailed material in the area of human reliability assessment. (Sources of human performance data for systems reliability determinations will be included in Chapter 6.) The final sections of this chapter will consider how human reliability may be enhanced.

What is human error?

Errors are defined in the *Concise Oxford Dictionary* as 'mistakes, conditions of erring in opinion or conduct'. They are a common feature of everyday life, both at work or outside. Rigby (1970) has usefully placed this definition into a working context and termed human error as 'any one set of human actions that exceed some limit of acceptability'. This approach begins to offer a systems definition of error. It suggests the need for setting performance standards based on analysis of work behaviour against systems criteria.

Human errors are not always the result of people being careless or inattentive (or 'bloody minded') and the average worker cannot easily be labelled as either accident- or error-prone or accident-free (Hale and Glendon, 1987). Indeed, making errors may be an important part of learning and maintaining skills, developing rules and problem solving (Rasmussen, 1986). Errors occur for many different reasons (for example, misperception, faulty information processing, poor decision making, and inappropriate behaviour) and under differing circumstances. They usually involve a complex interaction of many different factors.

Classification of error type or form (Table 5.1(a)) and error causation (Table 5.1(b)) can provide us with a valuable starting point for understanding error in terms of individual cognition. Numerous error taxonomies have been proposed. The HSE (1989) scheme detailed in Tables 5.1(a) and 5.1(b) is based on practical experience in the investigation of workplace accidents. Norman (1988) highlights two fundamental categories of error, 'slips' and 'mistakes'. He contends that slips occur as a result of automatic and routine actions under subconscious control, whereas mistakes result from conscious deliberations. This distribution can be mapped onto the taxonomy proposed by the HSE (1989) ('slips' encompass categories 3 and 4 in Table 5.1(a) and 'mistakes' categories 1 and 2). Errors are also often classified as being either errors of omission or commission (Swain, 1963). Errors of omission occur when the worker fails to perform some necessary action and errors of commission occur when an action is performed, but in an incorrect manner. This distinction could be viewed as an extension of category 4 in the HSE (1989) taxonomy.

Other taxonomies characterize errors in terms of the cognitive processes and behaviours generating them (Payne and Altman, 1962):

1. Input errors, errors of sensory input or perceptual processes.
2. Mediation errors, errors of mediation or information processing.
3. Output errors, errors in making physical responses.

This taxonomy maps onto the information-processing model of individuals described in Chapter 7.

Meister (1971) places errors in context and proposes that consideration should be given to the various stages of systems development and operation in categorizing error. He contends that certain types of errors are more likely during the design, as compared to the operation, of a system.

In summary, many taxonomies propose context-free explanations of errors *in general*, and virtually, by their very nature, couch those explanations in terms of individual psychological processes and behaviour (see, for example, HSE, 1989; Payne and Altman, 1962; Norman, 1988). However, concern for the reliability of specific systems (and more generally for the management of safety) requires that errors be classified (and explained) in the context of work processes (exemplified in Rasmussen, 1986 and Meister, 1971). Such context-dependent taxonomies have to consider the role and interplay of tasks, technological and organizational, as well as individual processes. Mapping taxonomies of error onto work processes opens up the possibility that errors might usefully be viewed in different ways in relation to different aspects of work: design of systems, implementation,

Table 5.1(a) Error type or form

	Type of error	
1.	Misperception	Preconceived ideas, poor judgement, no sound basis for judgement
2.	Mistaken priorities	Irreconcilable goals or objectives 'safety last!'
3.	Attention lapse	Incorrect course of action selected, performance error
4.	Mistaken action (includes errors of omission and commission)	Wrong actions (behaviours) performed impression right
5.	Wilfulness, violations, sabotage	Blatant disregard of procedures

Source: HSE 1989

Table 5.1(b) Error causation

	Cause of error	Example
1.	Inadequate information (knowledge and skill)	Inexperience, lack of confidence, resulting in poor judgement
		Poor training, ignorance of procedures and legislative constraints
		Communication failure, shift handover, badly documented procedures, lack of instruction and supervision, stand-ins
2.	Inadequate design	Plant not designed to 'fail-safe' mode
		Non-ergonomic work stations and equipment
		Poor environment, including inadequate ventilation, lighting and sound insulation
		Absence of systems of work, permit to work systems, vessel entry procedures etc.
3.	Individual differences	Inadequate selection, age, anthropometrics, gender etc.
		Memory capacity, decision-making, personality, mood, attitude, 'state of mind'
4.	Organizational culture	Safety 'last'

Source: HSE (1989)

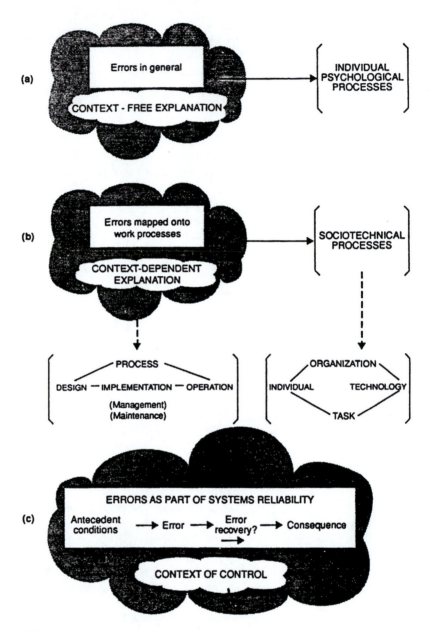

FIGURE 5.1
The basis of error taxonomies

operations, management and maintenance. Finally, when error is treated as a factor in systems reliability it may need to be seen as a process in itself, that of causation-kind-effect and recovery. This chain may need to be represented in whatever taxonomy is adopted. These three views of error taxonomies are presented in Figure 5.1.

Errors in routine behaviour

The different types of error categories discussed above have been the focus of much of Reason's work into the psychology of human error and he has sought to explain not only *what* errors can occur but also *why*. He has done so in terms of individual cognitive (mental) processes. Reason has argued that much of everyday performance and behaviour, both at home and at work, depends on the exercise of skills acquired through learning and perfected through practice (Reason and Mycielska, 1982). Such behaviour often takes the form of a sequence of skilled acts interwoven into well-established routines (see Chapter 7); the initiation of each subsequent act being dependent on the successful completion of the previous one. Errors in routine behaviours may result from both the structure and similarities of separate routines. In practice, events in the wider environment may automatically trigger an unintended action and produce either a replacement or a blend of routines (see Table 5.2). Norman (1988) has described these as 'capture errors'. One of the contributory factors in the Tenerife Air Disaster (Spanish Ministry of Transport and Communications, 1978) illustrates this phenomenon. The KLM captain took off without waiting for Air Traffic Control clearance. In his previous working role, as a trainer in a simulator, the captain had been used to giving his own clearance and in the adverse weather conditions at Tenerife analysts suggest that he may have reverted to his 'previous' behaviours to produce a new and totally inappropriate routine.

Table 5.2 summarizes Reason's explanation of errors in routine behaviour and provides an everyday example to illustrate the key points. Although errors involved in tea-making may seem trivial in comparison to errors involved in major disasters, we can translate Reason's explanation into everyday workplace situations and consider the cycle of routines which make up a short-cycle repetitive manu-facturing task. If an operator omits one of the key actions in the cycle or forgets where he or she is in the sequence of events, or if they mix different assembly tasks, then errors will occur. In this example, other workers may become involved if the task is part of a production line process or is dependent on learned performance.

Knowledge of *why* errors occur in well-practised behaviour is, in a way, more powerful than knowing *what* errors occur, and can be used in the design of the relevant systems and training of those that have to use them. The use of mimic diagrams for work-sequencing can be used to remind the individual where they are in a routine. Training sessions can be structured to reinforce key points and work can be organized to minimize distractions.

Errors in new behaviour

Often routine behaviour has to be overridden, and other 'newer' behaviours have to be acted out. This requires both the person's attention and their conscious control. However, errors can also occur

Table 5.2. Summary of theoretical framework of error in routine behaviour: errors in tea making (Reason and Mycielska, 1982)

	Error source		*Example*
1.	Individual selects wrong routine	1.	Selects coffee making routine instead of tea-making routine
2.	Stronger (better established) routine replaces intended one	2.	Coffee making routine is much more established and used and is therefore selected
3.	Individual omits what is a key action in the routine	3.	Omits to switch on the kettle and thus omits a key function in the routine
4.	Individual loses their place in the sequence, and either (a) jumps ahead, omitting a key action or, (b) repeats a completed action	4.	(a) Forgets to put tea in the tea pot (omits a key action), or (b) fills the pot twice
5.	Two actions compete for the next step in the routine and produce a curious blend of actions	5.	The telephone rings and the individual picks up the teapot and says 'hello'
6.	Individual forgets their intention and the sequence stops	6.	Walks to the cupboard to collect the mugs and forgets the tea-making altogether
7.	Intention is correct but the object is not	7.	Fills the sugar bowl with tea and not the mugs

here if the person is distracted or attends to the wrong aspect of the situation. In the first case, more established but inappropriate routines may immediately replace the new behaviour, and in the second, control over behaviour may be ineffective and the new behaviour fails under adverse conditions (stress or fatigue). The whole process of 'unlearning' should be carefully controlled and provides an example of how the context of the system can contribute predictably to the output.

Errors in complex behaviours

The increasing complexity of technical plant makes it difficult for operators to fully understand the system's total functions. Their workplace behaviours reflect the complexity of the operation which often requires them to perform a variety of separate tasks and the potential for error increases accordingly. Operators are particularly vulnerable to error during learning and adaptation (Rasmussen, 1986). They are also subject to errors of task interference. In a real life work situation the requirements from several different tasks will be considered by an operator on a time-sharing basis. Performance will thus be sensitive to the interactions between the various tasks and the level of mental activity (see later) required by the separate tasks.

In some cases, the operator may not have the mental resources to solve complex problems as they occur (for example an unplanned reaction in a chemical or nuclear process) and the consequences can be disastrous (see Chapter 17).

Errors as trigger events

Classification of error and error causation and theoretical frameworks of why errors occur may enable us to predict 'trigger' events. They support a failure mode and effect analysis of human reliability. One of the most common causes of errors is simply a lack of appropriate knowledge or skill, or inappropriate or inadequate instructions. These are closely followed by inadequacies in design (see Table 5.1(b)). 'Man–machine' systems which incorporate untrained people and are badly designed may therefore be particularly vulnerable (see Kletz, 1985). Swain and Guttman (1983) have termed these working conditions and system states as 'error-likely situations'. They are characterized by the mismatch between the operator's skills and capabilities and the demands of the system. However, even when the behaviour or action is well-known and well-practised, errors can and do still occur and the focus should be on the control of the effects of errors rather than on their elimination.

Characteristics of human error

Three extremely important aspects of human error are its obviousness (for self detection or detection by another person), its ability to be corrected (recovery) and its consequences (see Figure 5.2).

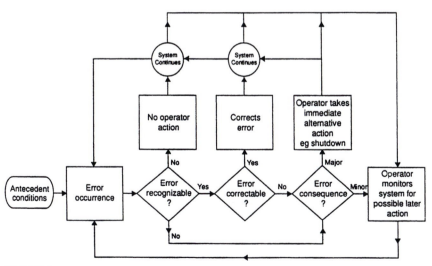

FIGURE 5.2
Flow diagram illustrating a possible error recovery sequence

If an error is made it may or may not be recognized (detected) by the operator as an error. If it is not recognized then the operator cannot take any further action. If it is recognized it may or may not be correctable (recoverable). In most cases, operators are motivated to take action to correct errors if they are able, but if the error is not correctable the consequences, which may be major or minor, must be considered. Major consequences would require the operator to take immediate alternative action, but minor consequences may only require continued monitoring to see if further action is required at a later time.

Modes of error detection

There are three ways in which people's errors are brought to their attention. Most directly, they can find out for themselves through various kinds of self-monitoring (or feedback). Second, something in the environment makes it very clear that they have made an error. The most unambiguous way by which the environment can inform us of our mistake is to block our onward progress. For example, if we have not turned the appropriate keys in a door it will not open. Similarly, computer-based systems are often programmed to respond to human error to limit 'system-damage' (see Lewis and Norman, 1986). Finally, the error is discovered by another person who then tells them. At Three Mile Island (Kemeny, 1979) the operators wrongly diagnosed the state of the plant. This was only discovered two and a half hours into the incident when the shift supervisor of the oncoming shift noticed the misdiagnosis. Each of these three detection modes is reviewed in a recent book on human error (Reason, 1990).

Human reliability – a systems context

The previous section developed our understanding of human error and focused on individual behaviours. It is important to extend this understanding into a wider systems context and to consider some of the problems associated with human reliability determinations within man-machine systems.

Human and technical reliability

Every man–machine system (see Chapter 7) contains certain functions which are allocated to the person, and a failure to perform these functions correctly or within prescribed limits can lead to systems failure. Hagen and Mays (1981) have produced a systems definition of human error as 'a failure on the part of the human to perform a presented act (or the performance of a prohibited act) within specified limits of accuracy, sequence or time, which could result in damaged equipment and property or disruption of scheduled operations'. In practice, the successful system performance depends not only on the human component but on the reliability of all the components. Technical and machine reliability have been considered in previous chapters and are the basis of reliability engineering. Human (or task) reliability has been defined as the probability of error-free performance within a specified period of time (Park, 1987) and in a particular context. Numerically this is 1 minus the probability of any human error within a given period for a particular system (see later). In practice, system reliability assessments are dependent on meaningful combinations of the separate component reliabilities (Park, 1987).

Figure 5.3 graphically illustrates the relationship between human reliability (R_h) and machine reliability (R_m) (represented by the emboldened curves). Thus a R_m of 0.9 coupled with a R_h of 0.8 results in a systems reliability (R_s) of only $R_s = 0.72$: a multiplicative function, typical of series reliability (see Chapter 3). If the human reliability

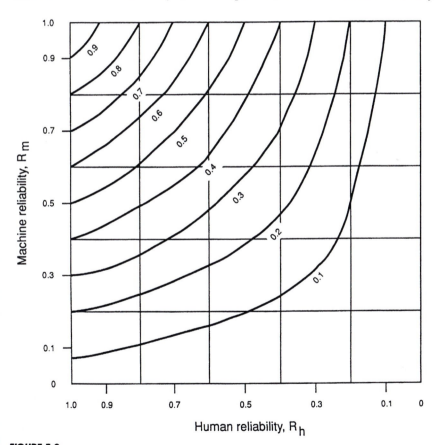

FIGURE 5.3
The relationship between human reliability (R_h) and machine reliability (R_m) (adapted from Park, 1987)

decreases to 0.6 with the machine reliability unchanged, the system reliability decreases to 0.54.

Many assessment models and procedures consider human reliability in terms of the individual operator's behaviour in relation to the relevant technology. Where many operators work as part of the same system, they are effectively treated as a non-interacting community of man–machine interfaces. This allows the application of the simple multiplicative function for describing overall reliability. Thus, if operator 1 has a reliability of 0.6 and operator 2 has a reliability of 0.7, then the overall reliability is 0.42. Such an approach, with its assumption of non-interdependence of reliabilities, fails to recognize the existence, importance and nature of team work. Where the team operates as the functional unit in a system, then either its reliability has to be assessed as different from that of the individual operators, or more sophisticated mathematical models of operator reliability have to be developed to account for the effects of team working (Cox et al., 1990). Such models need to take into account at least two observations. First, in team work, the probability of one person making an error is often dependent on the probability of other members erring and, second, there is the possibility that one person's error will be detected and corrected by another.

Nature of the task/task performance

In practice, human reliability analysts need to consider the nature of the tasks performed. This is usually done in terms of the demands it makes on the operator and the sequence of actions and behaviours. Early reliability methodologies were dominated by 'behaviourist' thinking, and measurements were taken of simple stimulus/response tasks to the exclusion of higher level decision making and problem solving tasks (Humphreys, 1988), and out of context of the overall system. The behaviourist view of the human as a mechanism (or a machine) fitted in with the way in which the human component was modelled in most systems reliability assessments. Probability data on required task perfor-mance were fed into conventional fault tree analysis in the same way as hardware component failure probabilities (see earlier). However, during the late 1970s the mechanistic approach to human reliability evaluation in systems began to be challenged, mainly due to the influence of Rasmussen (1986). Rasmussen reviewed a large number of incident and accident reports from nuclear power plants, chemical plants and aviation, and made the observation that 'operator errors' only made sense when they were classified in terms of the mental oper-ations being utilized in the task. His resulting skill, rule and knowledge (SRK) model of 'cognitive' control has become a market standard within the systems reliability community in assessing workplace tasks.

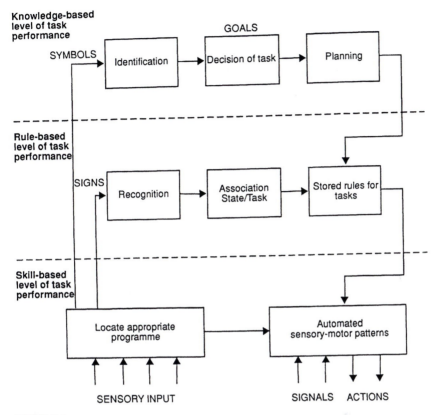

Knowledge-based level of task performance

Rule-based level of task performance

Skill-based level of task performance

FIGURE 5.4
Rasmussen's SRK framework of task performance

SRK framework

The framework relates to three distinct levels of task performance (see Figure 5.4). Each level relates to decreasing levels of familiarity with the environment or task. At the skill-based level, human performance is governed by stored patterns of preprogrammed instructions. It is characterized by 'free' and subconscious co-ordination between perception and motor actions. The rule-based level is applicable to tackling familiar problems in which solutions are governed by stored rules of the type:

IF   <STATE> THEN <DIAGNOSIS>
OR
IF   <STATE> THEN <REMEDIAL ACTION>

The knowledge-based level comes into play in novel situations for which actions must be planned on-line, using conscious analytical processes and stored knowledge.

With increasing expertise, the primary focus of control moves from knowledge-based towards skill-based levels; but all three levels can coexist at any one time. Skill-based behaviour is distinguished from rule-based behaviour by being far more resistant to outside interference (including stress) and for being most subject to built-in control of errors. Although a number of cognitive models have been developed which could be applied to human error modelling, Rasmussen's is probably the only model which has achieved widespread acceptance amongst human reliability assessors.

For example, Johannsen (1984) has used the Rasmussen model in fault management situations. He has argued that there are three main phases in fault management:

1. Fault detection
2. Fault diagnosis
3. Fault correction

Combining these three phases with the three cognitive levels proposed by Rasmussen gives nine possible categories of human operator task performance which enables investigators to better classify areas of interest for further assessment.

Performance level and error types

The SRK framework has been further extended (Reason, 1990) to form the basis of a generic error modelling system (GEMS). GEMS has been developed to provide a conceptual framework within which one may locate the origins of basic types of human error. It integrates two different areas of error research: slips and lapses, in which actions deviate from current intentions (Reason and Mycielska, 1982), and mistakes in which actions may run according to plan but where the plan is inadequate in some way (Rasmussen, 1986). GEMS yields three basic error types: skill-based slips, rule-based mistakes and knowledge-based mistakes.

Quantification of human reliability

Attempts to quantify human reliability have been incorporated into systems thinking since the late 1950s and originated in the aerospace industry. The majority of the work has taken place within those industries which are perceived as 'high risk' (for example, aerospace, chemical and nuclear process industry). It has been related to the probabilities of human error for critical functions and particularly in emergency situations (Humphreys, 1988). The most common measure of human reliability is the human error probability (HEP). It is the probability of an error occurring during a specified task. HEP is estimated from a ratio of errors committed to the total number of opportunities for error as follows:

$$HEP = \frac{Number\ of\ human\ errors}{Total\ number\ of\ opportunities\ for\ error}$$

The successful performance probability of a task (or the task reliability) can generally be expressed as $1 - HEP$.

Human reliability assessment techniques

The Safety and Reliability Directorate's *Human Reliability Assessors Guide* (1988) describes eight techniques for determining human reliability. The guide is written at a user level and provides detailed case studies for each of the following techniques:

1.   Absolute probability judgement (APJ);
2.   Paired comparison (PC);
3.   Tecnica empirica stima operatori (TESEO);
4.   Technique for human error rate prediction (THERP);
5.   Human error assessment and reduction technique (HEART);
6.   Influence diagram approach (IDA);
7.   Success likelihood index method (SLIM);
8.   Human cognitive reliability method (HCR).

It also provides detailed references for further information.

The majority of human reliability assessment techniques are based in part on behavioural psychology. They have been derived from empirical models using statistical inference but are not always adequately validated (Center for Chemical Process Safety, 1989). They are also dependent on expert judgements which may be subject to bias. Bias may be overcome by applications of techniques such as paired comparisons. This method does not require experts to make any quantitative assessments, rather the experts are asked to compare a set of pairs for which HEPs are required and for each pair must decide which has the higher likelihood of error (Humphreys, 1988). However, providing that the limitations of HEPs are recognized, it is often better to have carried through the process of deriving an 'acceptable' figure than to dismiss the task as impossible. Although it is important to note that such measurements should be applied with caution and in a way which takes account not only of the complexity of the overall system and of the potential accident process, but also of the exact nature of the task to which it refers. One of the most commonly used techniques for determining HEPs (THERP) is described below (Swain and Guttman, 1983).

Technique for human error rate prediction (THERP)

This technique, developed by Swain and Guttman (1983), can be broken down into a number of discreet stages which require the analyst to provide:

1. A system's description, including goal definition and functions;
2. A job and task analysis by personnel to identify likely error situations;
3. An estimation of the likelihood of each potential error, as well as the likelihood of its being undetected (taking account of performance shaping factors);
4. An estimate of the consequences of any undetected error;
5. Suggested and evaluated changes to the system in order to increase success probability.

Figure 5.5 provides a flow chart for THERP.

The original definition of the system's goals (see Chapter 1) is followed by a job and task analysis. In a task analysis the procedures for operating and maintaining the system are partitioned into individual tasks. Other relevant information (for example equipment acted upon, action required of personnel, and the limits of operator performance) is documented at this stage. Detailed information on this procedure is available in an Occupational Services publication (Patrick *et al.*, 1987). The task analysis is followed by the error identification process. The errors likely to be made in each task step are identified and non-significant errors (those with no important system consequences) are ignored.

This identification process should take account of the task models and error taxonomies discussed in earlier sections of the chapter.

The third stage is the development of an event tree. Each likely error is entered sequentially as the right limb in the binary branch of the event tree (see Figure 5.6).

The first potential error starts from the highest point of the tree at the top of the page. Each stage of the left limb thus represents the probability of success in the task step and each right limb represents its failure probability. To determine the probability of the task being performed without error, a complete success path through the event tree is followed. Once an error has been made on any task, the system is presumed to have failed unless that error is detected and corrected. The likelihood that an error will be detected and corrected must be taken into account by modifying the initial error probability.

The final stages are concerned with the assignment of error probabilities. Here the analyst estimates the probability of occurrence for each error, making use of all available data sources, formal data banks, expert judgements, etc. (see Chapter 6). Such estimates take into

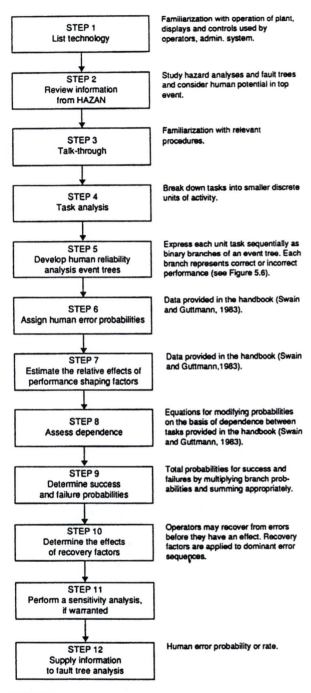

STEP 1 List technology	Familiarization with operation of plant, displays and controls used by operators, admin. system.
STEP 2 Review information from HAZAN	Study hazard analyses and fault trees and consider human potential in top event.
STEP 3 Talk-through	Familiarization with relevant procedures.
STEP 4 Task analysis	Break down tasks into smaller discrete units of activity.
STEP 5 Develop human reliability analysis event trees	Express each unit task sequentially as binary branches of an event tree. Each branch represents correct or incorrect performance (see Figure 5.6).
STEP 6 Assign human error probabilities	Data provided in the handbook (Swain and Guttmann, 1983).
STEP 7 Estimate the relative effects of performance shaping factors	Data provided in the handbook (Swain and Guttmann,1983).
STEP 8 Assess dependence	Equations for modifying probabilities on the basis of dependence between tasks provided in the handbook (Swain and Guttmann, 1983).
STEP 9 Determine success and failure probabilities	Total probabilities for success and failures by multiplying branch probabilities and summing appropriately.
STEP 10 Determine the effects of recovery factors	Operators may recover from errors before they have an effect. Recovery factors are applied to dominant error sequences.
STEP 11 Perform a sensitivity analysis, if warranted	
STEP 12 Supply information to fault tree analysis	Human error probability or rate.

FIGURE 5.5
A flow chart for THERP

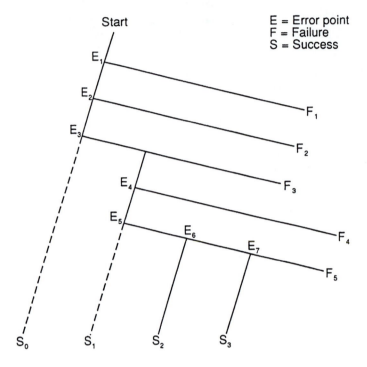

FIGURE 5.6
Event tree for human reliability

account the relative effects of performance-shaping factors (for example, stress, proficiency, experience) and all conditions which are assumed to affect task performance significantly. This process is often fairly arbitrary and may be one of the weakest steps in the procedure.

They also make an assessment of task dependence. Except for the first branch of the event tree, all branches represent conditional probabilities, with task/event interdependence directly affecting success/ failure probabilities. Thus, each task must be analysed to determine its degree of dependency. Each end point of an event tree is labelled as a task success or failure, qualified probabilistically and combined with other task probabilities to formulate total system success/failure probabilities. Details of mathematical calculations can be found in the *Human Reliability Assessors Guide* (Humphreys, 1988) or in the THERP handbook (Swain and Guttman, 1983).

Although, because of its mathematical basis, this method of error rate prediction implies accuracy, its utility is only as reliable and valid as the reliability and validity of its various measures. If these are not meaningful or themselves accurate then there is the real possibility that the overall process will itself be 'meaningless'.

Swain (1987) has also developed an annotated method as part of his accident sequence evaluation programme (ASEP). This method should be used only for initial screening and estimates the HEP for two separate stages in an accident or incident sequence:

1. Pre-accident;
2. Post-accident.

The pre-accident screening analysis is intended to identify those systems or subsystems that are vulnerable to human errors.

If the probability of system failure is judged to be acceptable using the method described in Table 5.3, human error is not important. If the probability of system failure is judged to be unacceptable, a specialist in the field of human reliability engineering should be consulted. Once an accident sequence has started, there is a chance that the operators will detect the problem and correct it before any serious consequences result (see Figure 5.2). For example, in a chemical process, the operators may detect that they have overfilled a reactor and drain off the excess reactant before heating the batch. If they fail to drain the reactor before heating, the reactor could be overpressured resulting in a release of toxic material. The post-accident human reliability analysis is intended to evaluate the probability of the operators detecting and correcting their error before the toxic material is released. Once the accident sequence has started, the most important variable is the time the operators have to detect and correct errors. Post-accident screening analysis provides this information.

Further details of the technique are available (Swain, 1987) and chemical process reliability examples which utilize such methods are given in a publication from the Center for Chemical Process Safety (1989).

Expert systems and human reliability analysis

Despite the development of annotated methods and techniques for human reliability analysis, such techniques continue to demand a fair degree of expertise. They rely heavily upon human factors/ergonomics analysts' judgement, particularly in the selection of appropriate information. In short, both the complex and annotated methodology are difficult for the non-human factors specialist.

There is a need for 'tools' to support and guide the non-expert in the selection and use of human reliability assessment. Computer-based technology is ideal for this purpose and an expert system (HERAX) has been developed at Aston University (Raafat and Abdouni, 1987). HERAX (Human Error Reliability Analysis eXpert) is written in common LISP and runs on an IBM (AT) PC and compatibles under the operating system DOS. The system is modelled on three of the previously listed techniques (THERP, SLIM and APJ). Although it has primarily been

Table 5.3 Pre-accident human reliability screening analysis procedure (adapted from Swain, 1987)

Step	Description
1.	Identify critical human actions that could result in accident
2.	Assume that the following basic conditions apply relative to each critical human action:

    (a) No indication of a human error will be highlighted in the control room

    (b) The activity subject to human error is not checked routinely (for example, in a post-operation test)

    (c) There is no possibility for the person to detect that he or she has made an error

    (d) Shift or daily checks and audits of the activity subject to human error are not made or are not effective

| 3. | Assign a human error probability of 0.03 to each critical activity |
| 4. | If two or more critical activities are required before an accident sequence can occur, assign a human error probability of 0.0009 for the entire sequence of activities. If these two or more critical activities involve two or more redundant safety systems (interlocks, relief valves, etc.), assign a human error probability of 0.03 for the entire sequence of activities |

designed for nuclear and process plant (Abdouni and Raafat, 1990) it can be adapted for other industrial and occupational situations.

Human performance data

Analysis of human performance and estimation of human error probabilities require supporting quantitative data. Objective data are derived from a number of sources, including laboratory studies, task simulators and operational observations (see Chapter 6). Error data may be presented in a number of forms (including time and frequency) and are produced by matching actual performance against an explicit or implicit set of requirements. Time measures obtained using instrumentation include reaction time and task duration. Frequency data are produced by counting numbers of operator responses, errors, outputs and events. The extent to which data obtained from one scenario (or type of analysis) can be generalized to others must be questioned. In the absence of such data, subjective or operator-based judgements have been used and a variety of psychometric techniques have been employed.

Numerous taxonomies of independent variables which affect human performance have been described in the scientific literature (see, for example, Tipper and Bayliss, 1987). An expert system (Human) has been developed which contains a rationalized taxonomy of independent variables. It was developed in a three-step process (Gawron *et al.*, 1989) including:

1. A review of existing taxonomies;
2. The addition of independent variables;
3. The removal of ambiguous and redundant data.

This database was produced as part of an ongoing development programme of a computer-aided engineering system for human factors engineers.

Human reliability enhancement

The concepts of human reliability discussed in previous sections can be usefully applied to enhance overall systems reliability.

Systems design

First, human reliability and its quantification is a central and important aspect of reliability and risk assessment (Watson and Oakes, 1988). It provides a means of determining the relative contribution of different sources of system failure. It is also essential for cost-benefit analysis to practise allocation of resources. Practical applications of human error analysis are described by Taylor (1988). A checklist similar to that depicted in Figure 5.7 may be incorporated into HAZOP procedures for each discrete step in chemical plant operating procedure (Taylor, 1988). Information from such analyses may then be incorporated into the final design features to enhance the reliability of the system.

Second, there is a need to design systems which can exploit the operator's ability not only to detect error but also to take corrective action. These objectives might be achieved by careful consideration of the way the operators' actions are registered by the system and whether the system displays 'ghost' those responses. The second possibility requires a delay in the system responding to the operator, and even being able to intelligently advise or help the operator correct the error. Essentially, we are concerned with designing 'forgiving systems', particularly in high- consequence environments. An example of a forgiving system is provided in the nuclear industry. If an operator mishandles the control rods in a British nuclear reactor then the reactor automatically fails safe and shuts down. In other words the engineering safeguards ensure safe operation even if the operator makes an error (Barnes, 1990). Similarly, procedures such as the '30-minute rule' enhance safety. They buy operators thinking time (for possible error recovery) in an emergency by demanding automatic systems capable of restoring the plant to a 'safe state' without the need for some human intervention during the first 30 minutes.

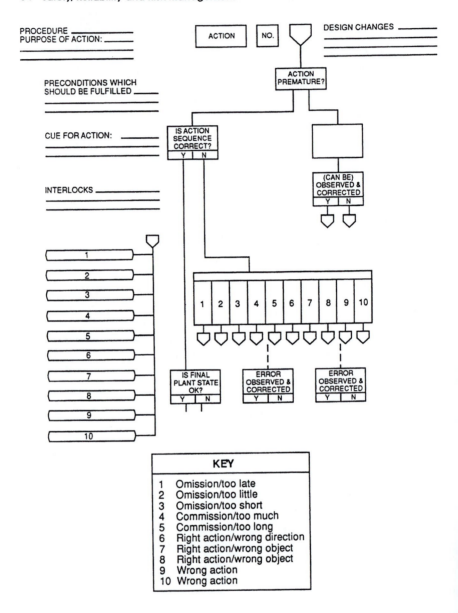

FIGURE 5.7
A human error analysis checklist

Transport systems (for example, rail systems) which are organized so that only minutes elapse between different commuter trains on the same line, may be unforgiving of driver error which breaches that brief window. Such systems may be protected by engineering safeguards, but these will not be infallible. Where these safety devices themselves might be under threat from possible human errors, it is necessary to build in independent back-up systems or 'redundancies'.

There is, therefore, a need to design systems so that there is a sensible balance between human and engineering control. Each needs to support the operation of the other. There may also be issues related to the organization of work as well as the design of the system and the behaviour of the individual. The Riso National Laboratory in Denmark (Rasmussen and Vicente, 1987) has produced guidance on

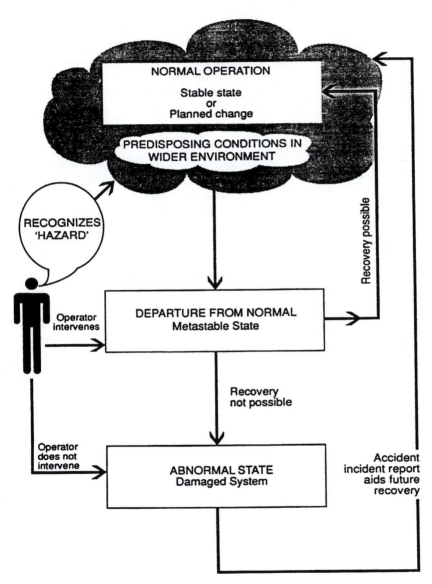

FIGURE 5.8
Accident sequence model

improved system design. The ten guidelines map onto the SRK model discussed in an earlier section and focus on increasing system tolerance to error.

Systems performance

Third, the operation and maintenance of 'safe' and reliable systems is obviously important. In this context information and understanding of human reliability are vital. It is also important that performance shaping factors such as stress and individual differences (described in Chapter 7) are taken into account in operational plans and that the role of the human operator within the system is fully understood. The mental models (see for example Goodstein *et al.*, 1988) which have attempted to link task performance and errors have facilitated this process.

The role of the human operator in accident prevention

We have stressed the importance of error detection and error recovery in the context of overall reliability. This is particularly important for errors with high consequence (for example, an error on a chemical process plant). Although engineering controls are essential, it is also important to recognize the *positive* role the human operator may play in responding to potential hazards and transient or metastable states in the system which were or were not directly attributable to their own errors.

Figure 5.8 presents an accident sequence model adapted from MacDonald (1972). It highlights the importance of processes of hazard recognition (see Chapter 7) and system recovery if stable states are to be maintained. THERP (Swain and Guttman, 1988) incorporates some assessment of human decision-making and action potential in responding to abnormal operating conditions and thus assesses the likelihood of human recovery. Such information is essential for the design and safe operation of systems and should be available to managers and designers.

Further enhancement of human reliability may be obtained from adequate selection, training and management of the human resource. This is discussed in later chapters.

Further reading

Goodstein, L. P., Andersen, H. B. and Olsen, S. E. (1988) *Tasks, Errors and Mental Models*, Taylor and Francis.

Reason, J. T. and Mycielska, K. (1982) *Absent Minded? The Psychology of Mental Lapses and Everyday Errors*, Prentice-Hall.

Reason, J. T. (1990), *Human Error*, Cambridge University Press, Cambridge.

References

Abdouni, A. H. and Raafat, H. M. N. (1990) Human reliability analysis in process industries and the application of expert systems technology, *J. Health Safety*, **4**, 5.

Barnes, M. (1990) *The Hinkley Point Public Enquiries*, Vol. 6, *The Risk of Accidents*.

Centre for Chemical Process Safety (1989) *Guidelines for Chemical Process Quantitative Risk Analysis*, The American Institute of Chemical Engineers.

Cox, S., Walton, C., Ferguson, E., *et al.* (1990) *Stress, Cognition and Control Room Operations*, Contract Report, Maxwell & Cox Associates, Sutton Coldfield.

Cox, S. and Cox, T. (1996) *Safety Systems and People*, Butterworth-Heinemann, Oxford.

Gawron, V. J., Drury, C. G., Czaja, S. J. and Wilkins, D. M. (1989) A taxonomy of independent variables affecting human performance, *Int. J. of Man-Machine Studies*, **31**, 643.

Goodstein, L. P., Andersen, H. B. and Olsen, S. E. (1988) *Tasks, Errors and Mental Models*, Taylor and Francis, London.

Hagen, E and Mays, G. (1981) Human factors engineering in the US nuclear arena, *Nuclear Safety*, **23**(3), 337.

Hale, A. R. and Glendon, A. I. (1987) *Individual Behaviour in the Control of Danger*, Elsevier, Amsterdam.

HSE (1985) *Deadly Maintenance: A Study of Fatal Accidents at Work*, HMSO, London.

HSE (1989) *Human Factors in Industrial Safety*, HMSO, London.

Hollnagel, E. (1993) *Human Reliability Analysis*, Academic Press, London.

Humphreys, P. (1988) *Human Reliability Assessors Guide*, United Kingdom Atomic Energy Authority, Warrington.

INPO (1985) *A Maintenance Analysis of Safety Significant Events*, Institute of Nuclear Power Operations, Atlanta, GA.

Johannsen, G. (1984) *Categories of Human Operator Behaviour in Fault Management Situations*. Proceedings of the 1983 International Conference on Systems, Man and Cybernetics, Bombay, India. December, p. 884.

Kemeny, J. G. (1979) *Report of the President's Commission on the accident at Three Mile Island*, Washington, DC.

Kletz, T. (1985) *What Went Wrong? Case Histories of Process Plant Disasters*, Gulf Publishing Co, Houston, TX.

Lewis, C. and Norman, D. A. (1986) Designing for error. In *User Centred System Design* (D. Norman and S. Draper, eds), Erlbaum, Hillsdale, NJ.

MacDonald, G. L. (1972) The involvement of tractor design in accidents, *Research Report* 3/72. Department of Mechanical Engineering, University of Queensland, St. Lucia.

Meister, D. (1971) *Human Factors, Theory and Practice*, Wiley, New York.

Norman, D. A. (1988) *The Psychology of Everyday Things*, Basic Books, New York.

Park, K. S. (1987) *Human Reliability Advances in Human Factors/Ergonomics*,

Elsevier, Amsterdam.

Patrick, J., Spurgeon, P. and Shepherd, A. (1987) *A Guide To Task Analysis Applications Of Hierarchical Methods*, Occupational Services, Aston, Birmingham.

Payne, D. and Altman, J. (1962) *An Index of Electronic Equipment Operability: Report of Development*, Report AIR-C-43-1162, American Institute for Research, Pittsburg.

Perrow, C. (1984) *Normal Accidents: Living with High Risk Technologies*, Basic Books, New York.

Raafat, H. M. N. (1983) The quantification of risk in system design, *J. Eng. Industry*, **105**, 223.

Raafat, H. M. N. and Abdouni, A. H. (1987) Development of an expert system for human reliability analysis, *Int. J. Occup. Accidents*, **9**(2), 137.

Rasmussen, J. (1986) *Information Processing and Human-Machine Interaction: An Approach to Cognitive Engineering*, Elsevier, Amsterdam.

Rasmussen, J. and Vicente, K. J. (1987) *Cognitive Control of Human Activities: Implications for Ecological Interface Design*. RISO-M-2660, Roskilde, Denmark: Riso National Laboratory.

Reason, J. T. (1989a) The Contribution of Latent Failures to the Breakdown of Complex Systems. Paper presented at The Royal Society, Discussion Meeting, Human Factors in High Risk Situations, London.

Reason, J. T. (1989b) Management Risk and Risk Management, Conference Proceedings, *Human Reliability in Nuclear Power*, Confederation of British Industry, London.

Reason, J. T. (1990) *Human Error*, Cambridge University Press, Cambridge.

Reason, J. T. and Mycielska, K. (1982) *Absent Minded? The Psychology of Mental Lapses and Everyday Errors*, Prentice-Hall, Englewood Cliffs, NJ.

Rigby, L. V. (1970) The nature of human error, in *Annual Technical Conference Transactions of the American Society for Quality Control*, Milwaukee, p. 457.

Rubinstein, T. and Mason, N. F. (1979) An analysis of Three Mile Island, *IEEE, Spectrum*, **37**.

Spanish Ministry of Transport and Communications (1978) Report of Collision between PAA B-747 and KLM B-747 at Tenerife, March 27, 1977, translated in Aviation and Space Technology, November 20.

Swain, A. D. (1963) *A method for performing a human factors reliability analysis*, Monograph SCR-685, Sandia National Laboratories, Albuquerque.

Swain, A. D. and Guttman, M. E. (1983) *Handbook of Human Reliability Analysis with Emphasis on Nuclear Power Plant Applications*, NUREG/CR-1275 (US Nuclear Regulatory Commission, Washington DC) 222.

Swain, A. D. (1987) *Accident Sequence Evaluating Procedure*. Human Reliability Analysis Procedure, Sandia National Laboratories, NUREG/ CR-4772 (US Nuclear Regulatory Commission, Washington DC).

Taylor, J. R. (1988) Using Cognitive Models to Make Plants Safe: Experimental and Practical Studies, in *Tasks, Errors and Mental Models* (L. P Goodstein, H. B. Andersen and S. E. Olsen, eds) Taylor and Francis, London, p. 233.

Tipper, S. and Bayliss, D. (1987) Individual differences in selective attention: The relation of primary and interference to cognitive failure, *Personality and Individual Difference*, **8**, 667.

Watson, I. A. and Oakes, F. (1988) *Management in High Risk Industries*, SARSS '88, Altrincham, Manchester.

Willis, H. (1962) *The Human Error Problem*, Paper presented at American Psychological Association Meeting, Martin-Denver Company, Denver.

Chapter 6

Data sources

Introduction

Large data banks providing information about component reliabilities, chemical properties, human reliability and incidents and accidents have been available for some years. The advent of small computers with convenient disk storage has made access very much easier and a number of large databases are now commercially available for 'desk top' use. This chapter restricts consideration to such large scale compilations, although much important information is to be found on a smaller scale in specialist books, journals and reports.

Mechanical and electrical component reliability data

In order to define the reliability of a component adequately it is necessary to make an accurate determination of the hazard rate or the reliability as a function of time for each failure mode under the specified operating conditions. This definition should then be clearly recorded along with the associated data. In practice, the reliability data available are likely to be based on information of poor statistical quality and may only provide mean failure rates as would be obtained if the only failure data recorded were the total numbers failing in a given time period.

There are many other difficulties. First, it is necessary to provide a clear definition of 'failure' for each mode. This may be obvious: a mechanical component fracturing or an electrical component going open circuit. On the other hand, failure may in some cases be defined on a sliding scale (for example, a heat exchanger efficiency dropping by 20% or an electrical resistor changing by 5%). The percentage adopted as the failure criterion for a particular component might well vary from application to application.

As previously discussed in Chapter 2, reliability can vary very rapidly with operating conditions. One very effective way of obtaining increased lifetimes is to de-rate components, while many external factors such as temperature, humidity and vibration can be very important. Thus, in quoting reliability data for a component, the operating conditions must ideally be accurately defined and some indication should, if possible, be provided of sensitivity to variations in these conditions. Reliability can also be profoundly affected by variations in quality control during manufacture, by storage and transport conditions prior to use, installation techniques and maintenance methods and standards. Thus, two batches or samples of apparently identical components may have very different reliabilities.

In specifying availability, further complications are encountered as replacement times, inevitably depending on accessibility, will need to be known.

Major reliability data banks

A large number of data banks are available worldwide. We concentrate here on two of the largest banks, one in the United Kingdom and one in the United States.

The AEA Technology Data Centre (AEA, 1996) holds reliability information on around 500 mechanical, electrical and control and instrumentation components and subsystems. Much of this information is in the main computerized data bank, but a large technical library holds much additional information in the form of handbooks and other documents. The holdings are summarized in Table 6.1. The Data Centre also provides access to a number of other sources including one of the main RAC databases from the United States.

Table 6.1 AEA Technology Data Centre – Summary of holdings

Component category	Number of components or sub-systems
Mechanical	218
Electrical	60
Control and instrumentation	142
Other	65
Total	485

The Reliability Analysis Centre (RAC, 1996) holds extensive electronic and non-electronic reliability data collections both in hard copy and on computer files. The main non-electrical data bank covers 1400 part types.

Chemical safety data

Chapter 4 discussed some of the chemical properties which need to be taken into account in the design of chemical plant to an adequate degree of safety and reliability. These properties are particularly associated with toxicity, flammability and explosiveness. Many tens of thousands of chemicals are in common use at the present time and the data banks needed to record the relevant chemical properties are correspondingly large.

In the United Kingdom, the introduction of the Control of Substances Hazardous to Health Regulations 1988 (COSHH, 1988) brought a new emphasis to the safe handling and use of hazardous chemicals. This, in its turn, has led to the more general use of chemical safety data in industry and elsewhere and manufacturers and suppliers are required to supply hazard data sheets with their products.

Major data banks

Commercial agreements have led to many of the data banks from worldwide sources becoming available on disk. For example, in the United Kingdom Silver Platter Information (Silver Platter, 1996) has issued CHEM-BANK which contains:

1. RTECS (Registry of Toxic Effects for Chemical Substances) from the United States, covering 120 000 chemicals.
2. HSDB (Hazardous Substances Databank) from the US Library of Medicine, covering 4500 chemicals.
3. IRIS from the US Environmental Protection Agency, containing risk assessment data.
4. CHRIS from the US Coastguard, containing information relevant to chemical spillage.
5. OHMTADS from the US Environmental Protection Agency with data on over 1400 materials designated as oil or hazardous materials.

An example of a hazard database issued by a chemical manufacturer is the 5000 item BDH data disk (BDH, 1996). This is available in six European languages.

The National Chemical Emergency Centre in Oxfordshire, UK, provides help and advice in dealing with chemical emergencies. The service is backed by a data bank (HAZDATA, 1996) containing information on 2800 substances.

The hazard index

The hazard index was first developed by the Dow Company in the United States (Dow, 1996) and was further developed in the United Kingdom by ICI Mond Division (Mond, 1996). The Dow or Mond Index takes into account not just the immediate properties of chemical or chemicals in use, but also the chemical processes which are taking place and the quantities of materials present. The resulting index is used as a guide to the safety features to be built into the process equipment.

The fire and explosion index contains a material factor multiplied successively by a general process hazard factor and a special process hazard factor. The material factor is normally evaluated for the most hazardous material present and depends on the flammability and the reactivity of the material. Material factors can also be defined for dusts having explosive potential. In such cases, the factor depends on the likely severity of explosion as defined by the maximum over-pressure and the maximum rate of rise of pressure.

The general process hazard factor depends on the general nature of the process. Exothermic reactions of various types, endothermic reactions, material handling processes and plant location all have bearing on this factor. The special hazard process factor depends on the process temperatures and pressures involved, material quantities, whether or not operation is in or near the flammable range and the likelihood of corrosion, erosion or leakage.

Many of the factors involved in the calculation of the Dow or Mond Index can be calculated from the properties of the materials involved, but the system makes the process very much easier by listing the factors directly. Many safety-related design features are determined with reference to the Index. These include siting and segregation, fire fighting requirements, electrical safety requirements, overpressure limitation requirements, spillage precautions and blast protection.

European inventory of existing commercial chemical substances (EINECS)

The Commission of the European Communities published the European Inventory of Existing Chemical Substances (EINECS) in the nine official languages of the Member States on 15 June 1990. This inventory, which had full legal effect from December 1990, lists those substances commercially available in the European Community over the past decade (a total of over 100 000). It does not cover medicinal products, narcotics, radioactive substances, foodstuffs or wastes. New products have to comply with the pre-registration conditions before they may be included on the update list ELINCS. EINECS and ELINCS are available on disk from Silver Platter (1996).

Human reliability (HR) data

The previous chapter has outlined the importance of the human component in systems reliability, this section considers HR databases and highlights some of the main problems associated with the collection of meaningful HR data.

Human reliability has been variously defined (see Chapter 5) and as an activity it is concerned with the analysis, prediction, and evaluation of work-created human reliability and performance in quantitative terms (Meister, 1971). HR values are derived using indices such as error likelihood, error probability, task accomplishment, and/ or response times. For example, a recent text on attention and performance (Warm, 1984) describes many experimental stimulus-response (S-R) studies using simulated vigilance tasks which provide empirical (S-R) data. However, systems reliability determinations on work-based tasks often require HR specialists to extend experimental data to real-life performance data and to be able to say that individual X or team Y will perform with a certain accuracy (probability of correct response of 0.98, for example). HR databases provide the basis of such predictions. Although many organizations may keep their own informal database there are 'formal' sources of data available to the HR practitioner or user.

Human error databases

There are only a few formal human error databases and the information they provide is limited. The primary database is the AIR Data Store (Munger *et al.*, 1962).

The Data Store is organized around common controls and displays (e.g. knobs, levers, meters). It consists of a compilation of performance data taken from 164 psychological studies (out of several thousand examined). It describes and fixes several characteristics of these controls and displays (for example, length of joystick). The data indicate the probability of successful operation of these instruments as a function of their design characteristics, together with an indication of minimum operation times. They also provide increments of time which must be added together when a component has multiple design characteristics. The goals of the technique include reliability predictions, identification of design features which degrade performance and guidelines for operator selection and training.

Various human reliability specialists have built on the Data Store to accommodate their specific needs. For example, Irwin and his co-workers developed the Data Store to predict personnel effectiveness during scheduled checkout and maintenance on the Titan II propulsion system.

Data Store and its scoring procedure have also been computerized. Further details of Data Store validity and reliability and its various applications are reviewed by Meister (1984).

A second source of human performance data is the Sandia Human Error Rate Bank (SHERB) (Humphreys, 1988). This is a compilation of Human Error Probabilities (HEPs) which are used in the technique for Human Error Rate Prediction (THERP) referred to in Chapter 5. This body of data consists of HEPs for many industrial tasks based on a large number of observations. Other human error databases are available and Topmiller *et al.* (1982) have performed a detailed review.

Judgmental task data

Where empirical data on tasks and task performance are not available, it is sometimes useful to use 'expert' estimates of reliability. This technique has been variously referred to, but is best known as, Absolute Probability Judgement (APJ). It relies on the utilization of 'experts' to estimate HEPs based on their knowledge and experience.

The method is described in Comer *et al.* (1984) and involves the following stages:

1. Selection of subject expert;
2. Preparation of task statements;
3. Preparation of response booklets;
4. Development of instructions for subjects;
5. Judgement;
6. Calculation of inter-judge consistency;
7. Aggregation of individual estimates;
8. Estimation of the uncertainty bands.

Two case studies utilizing APJ within the service sector and the offshore drilling sector are described in the *Human Reliability Assessors Guide* (Humphreys, 1988).

Human factors/engineering database

Engineers and designers have a special requirement for human factors data in designing reliable systems (see Chapter 7). An extensive and comprehensive engineering data compendium, *Human Perception and Performance,* has been published by the Advisory Group for Aerospace Research and Development (AGARD) (Boff and Lincoln,

1988). The compendium has been designed as a primary reference for system designers of human–system interfaces. It provides comprehensive information on the capabilities and limitations of the human operator, with special emphasis on these variables which affect the operator's ability to acquire, process and make use of task-critical information. It consists of concise two-page data entries of the following types:

1. Basic human performance data;
2. Section introductions outlining the scope of a group of entries and defining special terms;
3. Summary tables integrating data from related studies;
4. Descriptions of human perceptual phenomena;

5. Models and quantitative laws;
6. Principles and non-quantitative laws (non-precise formulations expressing important characteristics of perception and performance);
7. Tutorials on specific topics to help the user understand and evaluate the material in the compendium.

A similar computerized database (Human) was described in Chapter 5 (Gawron et al., 1989). This exemplifies the current trend to produce computerized data sources.

Accident and incident data

Various databases are available which provide valuable case histories. FACTS (TNO Division of Technology for Society) provides case histories of accidents with hazardous materials which happened worldwide over the last 30 years. It focuses on the following industrial activities: processing, storage, transhipment, transport and application. AEA Technology (AEA, 1996) have databases MHIDAS, EIDAS and EnvIDAS covering, respectively, major hazard, explosives and environmental incidents.

MARS

The major accident reporting system (MARS) was established in 1982 by the 'Seveso' Directive EEC/501/82. Under this Directive, Member States must notify major accidents to the EU for analysis and registration on MARS. A report has been issued covering 178 accidents in a ten-year period (EC, 1996).

The quality of accident information in MARS is usually accurate and is fairly extensive, and it has facilitated an exchange of official data on major accidents throughout the community.

Conclusions

All the databases discussed in the previous sections are useful for reliability determinations. The importance of good quality data cannot be overstated. However, great care must be taken to select data that are relevant to the situation under consideration if realistic predictions are to be made.

References

AEA (1996) AEA Technology Data Centre, Thomson House, Risley, Warrington, Cheshire, WA3 6AT, UK.

BDH (1996) Issued by Merck Ltd, Hunter Boulevard, Magna Park, Lutterworth, Leicestershire, LX17 4XN, UK.

Boff, K. R. and Lincoln, J. E. (1988) *Engineering Data Compendium, Human Perception and Performance*, Harry G. Armstrong Aerospace Medical Research Laboratory, Human Engineering Division, Wright-Patterson Air Force Base, OH 45433, USA.

Comer, M. K., Seaver, D. A., Stillwell, W. G. and Gaddy, C. D. (1984) *Generating Human Reliability Estimates Using Expert Judgement*, Vol. I, *Main Report*, NUREG CR-3688/10F2. 5 and 84-7115, GP-R-212022.

COSHH (1985) The Control of Substances Hazardous to Health Regulations 1988, HMSO, London.

Dow (1996) Dow Chemical Company, Midland, Michigan, USA.

EC (1996) European Commission – JRC, Public Relations and Publications Unit, I-21020, Ispra (VA).

FACTS, TNO Division of Technology for Society, P.O. Box 342, 7300 AM Apeldoem, The Netherlands.

Gawron, V. J., Drury, C. G., Czaja, S. J. and Wilkins, D. M. (1989) A taxonomy of independent variables affecting human performance, *Int. J. Man-Machine Studies*, **31**, 643.

HAZDATA (1996) National Chemical Emergency Centre, AEA Technology plc, Culham, Abingdon, Oxfordshire, OX14 3DB, UK.

Humphreys, P. (1988) *Human Reliability Assessors Guide*, Safety and Reliability Directorate, Wigshaw Lane, Culcheth, Warrington WA3 4NE, UK.

Meister, D. (1971) *Human Factors: Theory and Practice*, John Wiley, New York.

Meister, D. (1984) Human Reliability, in *Human Factors Review*, F. Muckler (ed.), Human Factors Society, Santa Monica, USA.

Mond (1996) Mond Index Services, 40 Moss Lane, Cuddington, Northwich, Cheshire, CW8 2PX, United Kingdom.

Munger, S., Smith, R. and Payne, D. (1962) *An Index of Electronic Equipment Operability: Data Store*, Report AIR-CA3-1/62-RP(1), American Institute for Research, Pittsburgh.

RAC (1996) Reliability Analysis Centre, Rome, NY, USA.

Silver Platter (1996) Issued by SCS, Room F15, Broad Oak Enterprise Village, Broad Oak Road, Sittingbourne, Kent, ME9 8AQ, UK.

Topmiller, D. Eckel, J. and Kozinsky, L. (1982) *Human Reliability Data Bank for Nuclear Power Plant Operators*, Vol. 1. *A Review of Existing Human Reliability Data Banks* (General Physics Corporation and Sandia National Laboratories) NUREG/CR-2744, USA Nuclear Regulatory Commission, Washington, USA.

Warm, J. S. (ed.) (1984) *Sustained Attention in Human Performance*, John Wiley, Chichester.

Chapter 7

Human factors in system design

Introduction

The UK Health and Safety Executive's (HSE) publication *Human Factors in Industrial Safety* (HSE, 1989) highlights the importance of human factors for improved health and safety within socio-technical systems. In the HSE's view, human factors encompass a wide range of issues, not only reflecting individual physical capabilities and mental processes, such as perception and cognition, but also person–environment interactions, equipment and systems design and the characteristics of organizations relating to safety (see Chapter 16). The framework used by the HSE to make sense of these different issues sets them in the context of the *individual* in their *job* within the *organization* (Cox and Cox, 1996) (see Figure 7.1).

Human factors' considerations, as defined by the HSE, are not restricted to issues directly related to people, but are also relevant to both technology and management systems and procedures. Two examples will suffice. First, the design of systems of management control must logically consider the structure and culture of the organization, the nature of the jobs covered, the abilities and characteristics of the staff involved, and the nature of the relations between management and those they are responsible for. Similarly, the design and introduction of safety technology, or of other engineering systems, must consider the ability and characteristics of the users (operators), their preparation (education and training) and a wide range of issues relating to the management of change set in the context of the organization.

Among other things, the HSE's human factors document (HSE, 1989) is seeking to promote a more integrated approach to health and safety. This involves bringing together engineering systems and controls of

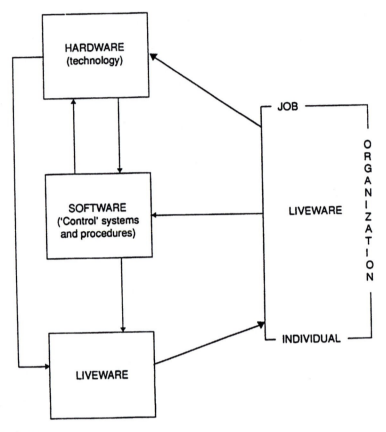

FIGURE 7.1
Human factors framework; individual, job, organization

plant and equipment (hardware), not only with efficient management systems and procedures (software) but also with a practical understanding of people and general knowledge of other human factor contributions. One powerful argument in favour of integrating these different areas of concern is the common observation that the majority of accidents are in some measure attributable to human as well as procedural and technological failure. For example, Hollnagel (1993) quotes figures which show a rise in human error rates from 25% of accident causation in the 1960s to 90% in the 1990s. This observation is supported by several HSE publications in the United Kingdom (HSE, 1985, 1987) and has been reported in other countries (for example, Dejoy, 1990, and Cohen *et al.*, 1975, in the United States).

This chapter will discuss three separate issues which typify current human factors concerns:

1. The nature of individual cognitive and physical capabilities;
2. The ergonomics of workstation design;

3.   The organizational context of socio-technical systems.

It will also consider the implications of each of these three areas for the design of safe systems. The reader will be referred to additional texts to extend their understanding and appreciation of the area. Case studies which highlight human factors considerations in major incidents will be included in Chapter 17.

The nature of individual cognitive and physical capabilities

How do we process information? Why do we sometimes see and hear things that are not there? How can we improve our everyday memory? What are attitudes? Why do we lose concentration when performing certain tasks? Why do we make mistakes? Why do we experience stress?

In answering these and other related questions, many psychologists explicitly adopt the systems approach discussed earlier in this book and treat the individual as an 'information processing system' (Cox and Cox, 1996; Hale and Glendon, 1987; Hollnagel, 1993). This computer analogy, which has aided our understanding of cognitive processes, makes a number of assumptions about:

- The separate stages of processing
- The selection and representation of information
- The limited capacity for information processing and how such limitations may be overcome
- The strategic management of information through the system.

It is worth noting, however, that individuals represent far more complex systems than most machines. When individuals are placed together in groups within an organization and form social systems, the interactions which naturally occur increase this complexity. Furthermore, individual cognitions and behaviours change in response to events or situations, many of which we, as 'observers', may not be aware of. Indeed, cognitions and behaviours may change when an individual is aware that they are being observed or are otherwise the focus of attention. Such effects add to the complexity of the overall system and to the sophistication necessary in any systems approach.

The individual as an information processor

Information has been defined formally as that which reduces uncertainty (Shannon and Weaver, 1949). Consistent with systems theory approach, Figure 7.2 presents the individual in terms of an information flow diagram. The arrows represent the flow of information through the system and the labelled boxes represent functional elements in the processing chain. Inputs into the system include various

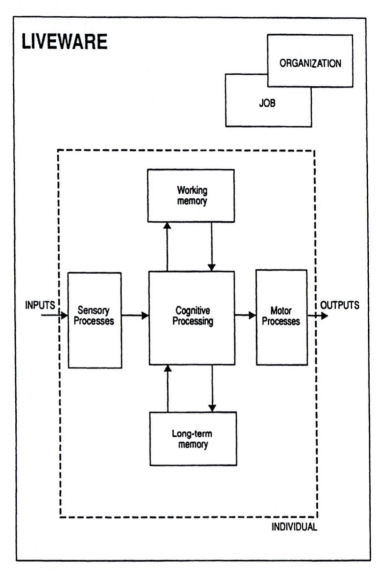

FIGURE 7.2
The individual as an information processor

sources of information and the outputs from the system are actions (behaviours). Figure 7.2 represents the basic human information processing (HIP) model which has been elaborated by authors such as Dodd and White (1980).

It may be constructive to work an example. Consider an operator's reaction to a visual alarm (visual stimulation). The significant stimulus is the flashing alarm light. The operator's sensory receptors would, in this case, be their eyes which would convert the flashing light stimuli into nerve impulses which would be transmitted via the optic nerve to the visual cortex and the central processing system. Here the information would be examined in the light of expectations and previous experience (by reference to memory) and a response of some kind would be selected. In this case, the response may be a decision to push down on a button with the right hand to stop the defaulting process. This decision on action would be coded as a series of nerve impulses which would be transmitted to the hand and arm muscles. These are the so-called motor processes. The motor processes convert the instructions into actions, which can be observed as outputs (or behaviours).

Key aspects of the information processing model will now be discussed in greater detail including:

- The sensory processes;
- Attention and perception;
- Cognitive processing and decision making;
- Motor processes and outputs.

There may also be feedback processes built into the alarm technology to reinforce a correct response or to inform a reconsideration and a second attempt. Information processing models thus imply at least some serial processing and some discrete processing functions. Although this approach has been criticized as simplistic (see Best, 1992) it offers a workable explanation of cognition and behaviour and is thus worthy of further consideration.

The sensory processes

Information is received by the human information processing system through the sense organs (see Table 7.1), the functional characteristics of which obviously place limits on the overall system. The threshold limits of senses are summarized, together with some common sensory defects, in Tables 7.2(a) and 7.2(b).

Table 7.1 The human senses

Organ	Sense
Eyes	Sight
Ears	Hearing/balance
Nose	Smell
Mouth	Taste
Skin	Touch, temperature, pain
Proprioceptors (from muscles or joints to brain)	'Kinaesthetic'

Table 7.2(a) Sensory thresholds

Sense modality	Threshold
Vision – energy	A candle flame seen at 30 miles on a dark, clear night
Vision – size	Detect an object 2 millionth of arc of circle (0.5 sec)
Hearing	The tick of a watch under quiet conditions at 20 feet
Taste	One teaspoon of sugar in two gallons of water
Smell	One drop of perfume diffused over the entire volume of a six-room apartment
Touch	Wing of a fly falling on a person's cheek from a distance of one centimetre

Table 7.2(b) Sensory defects

Sense	Natural and 'imposed' sensory defects
Sight	Colour blindness, astigmatism, long- and short-sightedness, monocular vision, cataracts, vision distortion by goggles and face screens
Hearing	Obstructed ear canal, perforated ear drum, middle ear damage, catarrh, ear plugs or muffs altering the sound reaching the ear
Taste and smell	Lack of sensitivity, genetic limitations, catarrh, breathing apparatus screening out smells
Touch senses	Severed nerves, genetic defects, lack of sensitivity through gloves and aprons

Hazards which are not perceptible to the senses, for example, x-rays, ultrasonics or gases such as methane, will not be detected unless suitable monitors and alarms are provided. Sensory defects may prevent information from arriving at the central processors, or distort it so as to make it unrecognizable or uninterpretable. In practice, defects can be caused by some of the equipment or clothing provided to protect people against exposure to danger. For example, an individual who is wearing safety spectacles has restricted peripheral vision and hearing defenders provided in 'noisy' environments may deprive wearers of vital auditory cues.

Not all the information which is available to the sensory processes (or detected by them) is 'used' by the person, and that which is used is first 'interpreted'. These processes of information selection and interpretation are known as 'attention' and 'perception'.

Attention and perception
The person is confronted with a vast array of different sources of information in the wider environment; only some of this information

they 'take in', interpret and use. Information is selected in two ways:

1. Peripherally by the nature and limitations of the person's sensory processes;
2. Centrally through cognitive mechanisms of attention.

At any time we are conscious of various things going on around us. In order to select what to attend to, we must subconsciously process a wider array of information rejecting much of it. This process could, in part, be peripheral but is more likely to be driven centrally, that is by cognitive processes.

An interesting example of selective attention is provided by the so-called 'cocktail party' phenomenon. At a party, amid all the noise and clamour, you can often concentrate on what one person is saying and 'cut out' all the rest. However, if somebody else mentions your name in some far corner of the party, your attention is suddenly drawn to them virtually as they speak. In order for this to happen, you must have been monitoring and processing, albeit at a low level, much of the information that was not reaching consciousness, remaining ready to 'switch back in'.

The information that we take in is usually incomplete, ambiguous and, at the same time, context dependent. If we simply 'saw' the world as it was projected onto our retinae then we would be very confused much of the time. We overcome this problem by 'interpreting' the available information, and actively building it into a mental model of our immediate world; this process of interpreting sensory information, etc., is referred to as 'perception'. Past experience may be a very powerful influence. There can be marked individual differences in perception and, on occasion, individuals may misrepresent reality. We may, for example, see what we expect to see rather than what is actually there.

Cognitive processing and decision making

The information that we attend to and then interpret contributes, with that stored in memory, to our mental model of the world (our 'knowledge' base). This model is the basis of our decision making. People make decisions in several different ways, and we know something about the processes involved and the rules they use (see Chapter 13). For example, information can be processed in two ways, either subconsciously or consciously. Subconscious processing is parallel and distributed. It occurs at many different places in the brain at the same time. It gives rise to what we recognize as 'intuition'. Conscious processing, by contrast, appears to be more logical and is serial in nature – a step-by-step process. Reason (1984) has referred to this process as being at the 'sharp end' of the information processing system (within the conscious work space). The conscious work space has a limited processing capacity. In both cases, some information is drawn from memory as much of the current information is incomplete. Indeed, what we currently 'know' of a problem may be altogether incomplete and our decision making has to work around this in various ways; for example, by using our knowledge of:

1.   Similar situations;
2.   Frequently occurring situations;
3.   Fecent situations.

An important part of the decision making process is therefore the ability to call on information which has been stored in earlier processing. This is what we mean when we refer to memory.

Memory

Memory contributes to all aspects of cognitive function, and as a system involves at least three processes:

1.   Encoding;
2.   Storage;
3.   Retrieval.

Information appears to be processed (or encoded) either verbally or iconically (by images) and storage may involve at least two sets of processes – short-term or working memory and long-term or permanent memory.

Short term memory processes appear to have the following characteristics:

1.   They have limited capacity;
2.   They lose information if it is not processed (cf. rehearsal);
3.   They support conscious processing;
4.   They involve *serial* processing.

One can increase storage capacity by (a) deliberately 'chunking' information in ever larger amounts, and (b) giving those chunks meaning. A chunk is an organized cognitive structure that can grow in size as more information is meaningfully integrated into it. For example, one can remember about seven single meaningless digits, but also about seven (ten digit) telephone numbers (Miller, 1956).

There is some debate over the existence of long-term memory. However, logically there is a need for some process by which the products for learning may be retained over the longer term. However, this process may be a natural continuation of the short-term or working memory. It appears to provide well-organized schemata or internal structures for organizing and retaining information. These schemata may be represented as a set of hierarchical structures in many ways similar to an active filing system (a changing system of files within files). Processing within these schemata is probably parallel (distributed) and unconscious. It gives rise to instinctive decisions, and is not necessarily logical but requires low effort. Existing schemata may change information as it is stored (adaptation) and may themselves change as that new information is incorporated (accommodation). Information is consolidated into these structures through use or importance.

Information may be retrieved from the memory process by one of two commonly studied processes: recall and recognition. These are different; for example, we can recognize somebody without being able to recall their name. Recognition is easier than recall because useful cues are obviously present in the person or thing being recognized. Recall can be improved with aids to recall such as mnemonics. Mnemonic devices are often incorporated into brand names, e.g. Easy-Off oven cleaner.

There are several different aids to memory and information processing which may be of importance in relation to safe designs:

1.   Redundancy of information (this needs to be balanced with the need to prioritize essential information);
2.   Minimal interference;
3.   Meaningfully chunked information;
4.   Flagging of important information;
5.   Use of mnemonic devices incorporated (recall aids);
6.   Minimal encoding requirements (dials say exactly what is meant by particular readings);
7.   Provision of attentional devices (for example, flashing lights or intermittent noise).

Information retrieved from memory contributes to decision making which may, in turn, result in action often.

Attitudes as framework for decisions

Attitudes provide an important framework within which decision making is made. They are relatively stable, but not unchangeable, components of the person's psychological make-up. They are developed through experience, and may be heavily influenced by cultural, sub-cultural and local social pressures. Attitudes can be defined in terms of mini belief systems or tendencies to act or react in a certain (consistent) manner when confronted with various (trigger) stimuli (Cox and Cox, 1996).

Explicit in many definitions of attitudes is the notion that they are involved in determining the way the person thinks, feels and behaves in relation to particular situations or events. Table 7.3 illustrates this approach using both positive and negative attitudes to safety.

Table 7.3 Attitudes to safety

Component	Positive attitude	Negative attitude
Thinking (cognitive)	X is aware of and thinks carefully about safety	X thinks safety is just common sense
Feeling (affective)	X is enthusiastic about safety	X is bored by talk of safety
Doing (conative)	X complies with safe working procedures	X ignores safe working practices

Limited capacity

The concept of 'limited capacity' is an important one for our model of the individual as 'an information processor'. It means that the system can be overloaded. It also provides the requirement for both selective attention and then the allocation of processing resources in order that a multitude of information processing tasks can be dealt with during one period of time. Naturally mistakes and errors can be made if the wrong information is attended to or the wrong tasks are given (processing) priority. Similarly, the limitations to memory processes may restrict our performance.

In addition to providing a framework for decision making, attitudes may also serve as 'filters' and contribute to the higher processes of attention. Information and messages will be more readily accepted if they are perceived as existing attitude and belief systems or act to reduce any inconsistencies in those systems.

Motor processes and outputs

The output of the system (see Figure 7.2) is some form of behaviour, which might be verbal (what they say) as well as locomotor (what they do). Such behaviour is the result of the various cognitive processes described in the previous sections. Interestingly, while people's actions (what they do) are the prime concern for safety, their verbal behaviour (what they say) can have an indirect but strong effect on the safety of the system. What people say contributes to the communication of ideas about, and attitudes to, safety, shapes expectations and can reward particular behaviours in others. The next section highlights some important aspects of human behaviour.

Human behaviour

Much of everyday behaviour, both at home and at work, depends on the exercise of skills acquired through learning and perfected through practice. Such behaviour often takes the form of a sequence of skilled acts interwoven into well-established routines. The initiation of each subsequent act being dependent on the successful completion of the previous one.

Learning and control

These routines are learnt, and during learning the person has to pay close attention to what they are doing, monitoring and correcting mistakes, often as they occur. This process is best described as a feedback system (Norman, 1988). However, as learning progresses and the sequence of skilled acts is perfected and strengthened then control over the sequence is delegated to the unconscious mind, and the lower centres of the brain. At this stage, the process becomes automatic and is now better described as an open system; once initiated the routine will run through to completion without any requirement for conscious control. The person no longer has to attend to what they are doing. This allows them to focus on other things, and process other sources of information. Automaticity confers real advantage. However, the new automatic system is not infallible and errors do occur. Indeed, the very act of thinking about a skilled routine may disrupt it, for example, if you consciously consider the stages in the 'walking downstairs' routine you will probably fall over (see Chapter 5). Similar errors can occur when a person is carrying out a routine task in the workplace.

Perception of risk

Our behaviour in any situation (hazardous or non-hazardous) is moderated by our perception of risk. When we make judgements about things in the environment, whether it be a simple structural characteristic such as the size of an object or an attribute like risk, we are swayed in those judgements by contextual information. This information can lead us to illogical conclusions. It is important to ensure that information on hazards is presented in a clear, unambiguous form and that we as individuals receive as much factual information as possible. (The perception of risk is covered in more detail in Chapter 13.)

Behaviour in hazardous situations

The aspects of cognition and behaviour described in the previous sections are all important when we consider human behaviour in hazardous situations. When an individual encounters a hazard or is confronted by a potentially hazardous situation, a number of activities take place (Dejoy, 1990) including:

- hazard perception;
- hazard cognition (i.e. a knowledge, awareness and an understanding of the hazard);
- the decision to avoid the particular hazard based on individual perceptions of the associated risk;
- exercising the necessary abilities to take appropriate action (self protective behaviour).

At each stage in the sequence the probability of an accident or systems failure either increases or decreases.

Systems designers should consider the implications of each of these stages for the design of safe systems. Similarly management should ensure that personnel are selected and trained in the use of equipment and have the necessary information and the required abilities to behave 'safely' (see later).

Dejoy (1990) has emphasized the importance of decision making for self-protection and has proposed a model for obtaining diagnostic information on such behaviours in working situations. In his model the inability to take appropriate action (i.e. failure to engage in self protective behaviour) is analysed in terms of three factors. These are later used as the basis for selection, prevention or intervention strategies.

Individual differences
Even when people are given the same information on hazards, their responses may vary. Such differences in behaviour may result from differences in perception, experience, attitude, personality or skill. Individuals differ in many ways, and many of these ways can be measured and their implications for safe behaviours studied. For example, the concept of 'accident proneness', originally described by Greenwood and Woods in 1919, has driven much of the research into individual differences and accident causation (see Sheehy and Chapman, 1987). Accident proneness has, however, not proved to be a useful concept in terms of predicting individual performance. However, there are accident repeaters and these individuals often have sensory or motor disadvantage with respect to their task or task environment.

A more promising approach, therefore, is to consider how best to match the person's characteristics, broadly and generally defined, to the demands of the task and task environment. It is work in this area of general concern that underpins selection.

Stress
The experience of stress by people in any socio-technical system can lead to behaviours which cause systems failure. Stress is a complex psychological state deriving from the person's cognitive appraisal (see Figure 7.3) of their failure to easily adapt to the demands of the environment (work or home) and the extent to which that environment meets their needs. It exists in the person's recognition of their inability to cope with the demands of the (work) situation and in their subsequent experience of discomfort. Stress is thus not an observable or discrete event. It is not a physical dimension of the environment, a particular piece of behaviour, or a pattern of physiological response.

It has been suggested that the process of appraisal takes account of at least four factors (Cox, T., 1985):

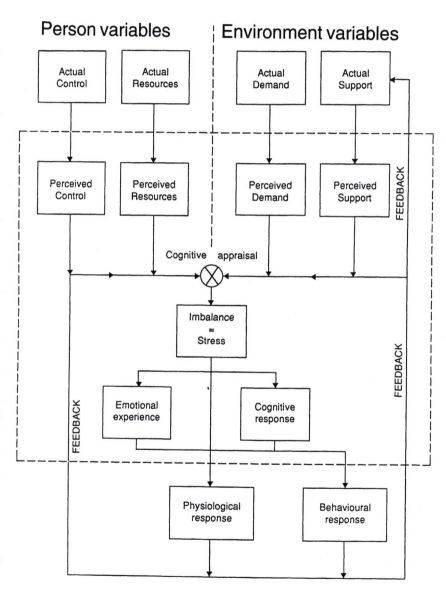

FIGURE 7.3
The transactional model of stress (adapted from Cox, 1978)

1. The demands on the person matched against:
2. Their ability to meet those demands (personal resources).
3. The constraints that they are under when coping.
4. The support received from others in coping.

The absolute level of demand would not appear to be *the* important factor in determining the experience of stress. More important is the *discrepancy* that exists between the level of demand and the person's ability to cope (personal resources). Within reasonable limits a stress state can arise through *overload* (demands>abilities) or through *underload* (demands<abilities). It has been added that a state of stress may exist only if the person believes that the discrepancy is significant.

It should be obvious that the notion of a stress state is different from that of an arousal continuum (wakefulness/alertness/vigour) and the two concepts should not be confused. Demand or challenge may be arousing but they do not necessarily produce a state of stress.

Effects of stress

There appears to be an immediate response to the perception of a stressful situation in the form of a negative emotional (unpleasant) experience. There is, however, no single diagnostic stress emotion; rather there is a variety and mixture of negative feelings probably reflecting individual disposition as well as situational factors. For an otherwise normal population, this reaction may be reflected in a general change in mood. This immediate response is of interest because it provides the person with a signal or criterion by which he or she can identify stressful situations and then monitor their own progress in dealing with them.

The emotional experience of stress is often accompanied by changes in the person's perceptual and cognitive processes and in behaviour and physiological function. Indeed, it is popular to categorize the responses to stress as psychological, behavioural or physiological (see Table 7.4). In the context of safety and reliability it is also important to consider the effects of stress on individual and organizational performance.

Some responses are more controlled and planned than others. Those that are deliberate attempts at mastering the problem situation or, more simply, dealing with the experience itself, are often termed coping.

The experience of stress may result from failures adequately to take into account people's needs and abilities in the design of tasks, technologies and the work environment. Consideration of person–job fit is one of the fundamental aspects of ergonomics.

Table 7.4 The effects of stress (adapted from Cox, 1978)

1. *Subjective*	2. *Behavioural*
Anxiety, aggression, apathy, boredom, depression, fatigue, frustration, guilt and shame, irritability and bad temper, moodiness, low self-esteem, threat and impaired tension, nervousness, loneliness and emptiness, restlessness, and trembling	Accident-proneness, drug taking, emotional outburst, excessive eating or loss of appetite, excessive drinking and smoking, excitability, impulsive behaviour, speech, shouting, nervous laughter
3. *Cognitive*	4. *Physiological*
Inability to make decisions and concentrate, frequent forgetfulness, hypersensitivity to criticism, and mental blocks	Increased blood and urine catecholamines and corticosteroids, increased blood glucose levels, increased heart rate and blood pressure, dryness of mouth, sweating, dilation of pupils, difficulty in breathing, hot and cold spells, 'a lump in the throat', numbness and tingling in part of the limbs and 'butterflies' in the stomach
5. *Health*	6. *Organizational*
Asthma, amenorrhea, chest and back pains, coronary heart disease, diarrhoea, faintness and dizziness, dyspepsia, frequent urination, headaches and migraine, neuroses, nightmares, insomnia, psychoses, psychosomatic disorder, diabetes mellitus, skin rash, ulcers, loss of sexual interest and weakness	Absenteeism, poor industrial relations and poor productivity, high accident and labour turnover rates, poor organizational climate, antagonism at work and job dissatisfaction

Ergonomics of workstation design

Traditional ergonomics is concerned with fitting tasks, technology and work environments to the known characteristics of the 'operator'. In its early years it was largely concerned with the design of the physical work environment, with issues surrounding displays and controls, workstation layout, seating, and heating, lighting, noise and ventilation, and with concepts such as population stereotypes and stimulus–response compatibility. More recently, it has become deeply involved in the design of computer programmes and the human–computer interface, and the term 'cognitive' ergonomics has been coined. Ergonomics is about fitting the job to the person and complements areas such as selection and training which seek to fit the person to the job.

Ergonomics has twin objectives: the maximization of performance and the optimization of well-being. Often the latter is given expression as a reduction in the human cost of performance. Information is readily available to support these objectives in the following areas:

1.   Standards and normative data relating to the person for use in design exercises (anthropometrics, see for example NASA, 1978; Kroemer, 1989);
2.   Procedures for investigating ergonomic issues (including surveys, see for example Table 7.5);
3.   Principles of good practice.

Together these can be used to ensure satisfactory system designs.

Failure to seriously consider ergonomic issues can lead not only to impairments of performance but also to ill health and accidents. Relatively dramatic examples of health effects of poor task and workplace designs exist in relation to the alleged reproductive health hazards of visual visplay units (VDUs) and the incidence of repetitive strain injury (RSI) in repetitive work (Cox, S., 1985).

The effects of poor task and workplace design can also disadvantage particular groups of workers (for example, those who are small or tall, relatively weak, etc). Those most 'at risk' of disadvantage differ from the 'average' person in some particular way. Equally, the application of ergonomic procedures and principles can reduce the impact of individual differences at work. Fully adjustable workstations and seats take out differences in size and reach. Well designed tools (levers, etc.) can minimize the need to be 'strong', and so on.

Person–machine interaction

Part of the process of fitting the job to the person may be a consideration of the person–machine interface (PMI). Inherent in this interaction process is the transfer of either energy (power) or information between the person and the machine. This exchange takes place via an imaginary plane known as the 'person–machine interface'. This information passes from the machine to the person through the display elements of the interface and from the person to the machine through the so-called control elements of the interface. Here the person and the machine are combined in a closed-loop feedback system (see Figure 7.4).

Another concept fundamental to person–machine system ergonomics is that of allocation of function between the person and the machine. The problem can be seen as one of defining the functional location of the interface depicted in Figure 7.4. The solution is not simply one of separating functions appropriate to the person or the machine but must also consider the social, economic and political context to that problem. Thus, the introduction of computerized and automated production systems may have implications for de-skilling and levels of employment

Table 7.5 Human factors/ergonomics survey descriptor list (adapted from Eastman Kodak Company, 1983)

Workplace characteristics and accessories (equipment)
Reaches
Clearances
Crowding
Postures required
Chairs and footrests
Heights
Location of controls and displays
Motion efficiency
Workplace accessibility (as in moving supplies into it)

Environment/physical
Noise level and type
Vibration level
Temperature
Humidity
Air velocity/dust and fibres
Lighting quantity
Lighting quality, especially glare
Electric shock potential
Floor characteristics, including slipperiness, slope, smoothness
Housekeeping
Hot and cold surfaces
Protective clothing needed

Physical demands
Heavy lifting or force exertion
Static muscle loading
Endurance requirements
Work–rest patterns
Frequency of handling
Repetitiveness
Grasping requirements
Size of articles to be handled: very large or very small
Sudden movements
Stair or ladder use
Tool use

Environment/mental
Skill requirements
Multiple tasks done simultaneously
Pacing
Training time needed
Monotony: low challenge
Concentration requirements
Information demands including processing
Complexity of decision making, defect recognition

Displays, controls, dials
Size/shape relative to viewing distance
Compatibility
Display lighting
Labelling
Internal consistency

Perceptual load
Visual acuity needs
Colour vision needs
Space and depth perception requirements
Tactile requirements
Darkroom vision
Auditory demands
Stress

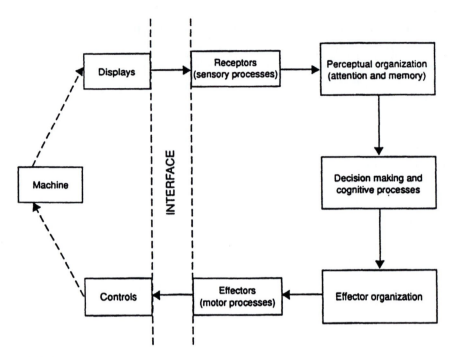

FIGURE 7.4
The man–machine interface

which go far beyond consideration of an individual operator's role in the work process. Nevertheless, in the design of most processes where the human operator is involved, the problem of allocating functions must be resolved.

On what basis should this task allocation be decided? The most obvious way to separate functions, given that constraints such as cost–benefit have been considered, is on the relative capabilities and limitations of the two components. Such an approach was attempted by Fitts (1951) in drawing up his now famous list itemizing the relative advantages of men and machines. An updated version of the so-called 'Fitts List' as modified by Singleton (1974) is shown in Table 7.6.

It is clear from an examination of this list that humans appear to surpass machines in detection, pattern recognition, flexibility, inductive reasoning and judgement. On the other hand, machines appear to surpass humans in speed, precision of response and application of sustained power, repetitive performance, short-term memory, deductive reasoning and multi-channel performance. The human does, however, have one other particular crucial ability: that of an in-built error-correction and error-monitoring facility which, coupled with great flexibility and versatility, means that he or she can often detect, then act to minimize the consequence of error (see Chapter 5).

However, the 'Fitts List' approach has at least four main disadvantages:

Table 7.6 An updated version of the Fitts List (from Singleton, 1974)

Property	Machine performance	Human performance
Speed	Much superior Consistent at any level Large constant standard forces and power available	Lag 1 second 2 horse-power for about 10 seconds 0.5 horse-power for a few minutes 0.2 horse-power for continuous work over a day
Consistency	Ideal for routine, repetition and precision	Not reliable – should be monitored Subject to learning and fatigue
Complex activities	Multi-channel	Single channel Low information throughout
Memory	Best for literal reproduction and short-term storage	Large store multiple access Better for principles and strategies
Reasoning	Good deductive Tedious to reprogramme	Good inductive Easy to reprogramme
Computation	Fast, accurate Poor at error correction	Slow Subject to error Good at error correction
Input	Can detect features outside range of human abilities Insensitive to extraneous stimuli Poor pattern detection	Wide range (10^{12}) and variety of stimuli dealt with by one unit, e.g. eye deals with relative location, movement and colour Affected by heat, cold, noise and vibration Good pattern detection Can detect very low signals Can detect signal in high noise levels
Overload reliability	Sudden breakdown	Graceful degradation
Intelligence	None Incapable of goal switching or strategy switching without direction	Can deal with unpredicted and unpredictable Can anticipate Can adapt
Manipulative abilities	Specific	Great versatility and mobility

1.   It tends to become quickly outdated with the current rate of development in areas such as micro-electronics technology;
2.   It can only offer a rough guide in the first stages of design and should be modified towards the final stages of the process;
3.   No adequate systematic methodology exists in which highly quantified engineering data can be reliably contrasted with comparable data on human performance (see Chapter 6);
4.   Allocation of function should allow the individuals an opportunity not only to utilize their existing skills but also to develop these. Such an approach is the basis of person-centred ergonomics described by Oborne in his book 'Ergonomics at Work' (Oborne, 1995).

Design of person–machine systems should also take into account population stereotypes.

Person-centred ergonomics

Traditional PMI approaches now face a further challenge. Oborne has argued in his book *Ergonomics at Work* (Oborne, 1995) that, although individuals and their working systems should operate in close harmony, in practice, operators and 'working-systems' are not equal partners. Furthermore, he considers that debates on equality not only denigrate the most important component in person–machine systems (people) but may also be considered to reduce them to the level of inanimate components. Thus, the 'person-centred' view of ergonomics argues for 'person' control of systems. Modern ergonomists further propose that this view points to an approach to system design from the primary standpoint of the operator rather than the traditional 'person–machine' perspective. This 'person-centred' approach firmly puts people at the centre and points up the following essential features for the design of safe and effective systems:

1.   *Purposivity* – the technology needs to reflect the actual use to which it is put (not the perceived use);
2.   *Anticipation and prediction* – these follow-on from the concept of purpose; they concern how the system is operated and controlled. For example, the way in which information is displayed to an operator should be such that they can 'see' the results of their actions before they are carried out (see, for example, the Three Mile Island incident described in Kemeny, 1979);
3.   *Interest and boredom* – this feature relates to the stimulation and interest of the operator and stems from the source of the activity. Increased interest leads to a lowered likelihood of boredom and subsequent reduction in errors;

4.   *Control and autonomy* – the importance of these concepts is well-recognized in organizational psychology – control (real or perceived) over the situation is paramount and reduces the uncertainty of the outcome;

5.   *Responsibility and trust* – a central aspect of the person-centred approach is that individuals act with responsibility when interacting with the system. Since this responsibility is towards the successful outcome of the goals the information must be of the kind and nature necessary to facilitate the desired outcome. Any information received must also be trusted by the operator.

Population stereotypes

People expect things to behave in certain ways when they are operating controls or when they are in certain environments. Although it is possible to educate people to operate systems that do not follow the stereotypes, their performance may deteriorate when placed in an emergency situation.

Some examples of stereotypes are provided in *Ergonomic Design for People at Work* (Eastman Kodak Company, 1983). Some of these examples have implications for workplace safety and should be considered at the design stage:

1.   Very loud sounds, or sounds repeated in rapid succession, and visual displays that blink or are very bright imply urgency and excitement;

2.   Seat heights are expected to be at least 40 cm (15.5 in) above the floor in production workplaces and offices;

3.   Very large or dark objects imply heaviness. Small or light-coloured objects imply lightness. Large, heavy objects are expected to be at the bottom and small, light ones at the top;

4.   Red signifies 'stop' or 'danger', yellow indicates 'caution', green indicates 'go' or 'on', and a flashing blue indicates an emergency control vehicle, such as a police car;

5.   Knobs on electrical equipment are expected to turn clockwise for 'on', to increase current, and counter-clockwise for 'off', to decrease current;

6.   Wheels or cranks to control direction of a moving vehicle are expected to use clockwise rotation to make a right turn and counter-clockwise rotation to make a left turn;

7.   For vertical levers that move in the horizontal plane (e.g., crane controls), movement away from the body is associated with decreasing action (lowering) and movement toward the body with increasing action (raising). Movement of a lever to the left should be associated with movement of the object controlled to the left also;

8. Pulling a control such as a throttle outward from a panel signifies that it has been activated (on). Pushing it in disengages it (off);

9. For controls mounted overhead (e.g. on the ceiling of a control booth), pushing forward (away from the body) specifies increasing (on) activity, and pulling back specifies decreasing (off) activity.

Each of these examples demonstrates the importance of this area of ergonomics for systems design. However, the designer of workstations and jobs should also consider the effect of the wider 'social' environment on individual performance. This involves an understanding of the organization.

Organizational context of sociotechnical systems

For the majority of people, work takes place *within* an organizational context and most of those people think of themselves as working *for* organizations (Cox, Leather and Cox, 1990). It is therefore important for designers to examine the relationship between the key features of organizations and those of jobs and the resulting workplace behaviours of individuals.

In highlighting the key features of organizations we must begin by asking the most basic question, 'what is an organization?'. David Buchanan and Andrzej Huczynski (1985) define organizations as 'social arrangements for the controlled performance of collective goals'.

There are three elements to this definition:

1. That organizations are *social arrangements*: organizations are about groups of people interacting with each other in particular ways;

2. That organizations are concerned with achieving *collective goals*: members of organizations share at least some common goals;

3. That organizations survive through *controlling performance*: organizations are concerned with performance in pursuit of their goals and that such performance is controlled through a variety of means from training and the exercise of management authority to the way work and jobs are actually designed.

The purpose of designing organizations is therefore to develop management influence on individual behaviour in order to solve the basic problems of task achievement, controlled performance and cost-effectiveness.

Organizations, safety and reliability

It is easy, though misleading, to think of the organization only as the embodiment of its plant, machinery and personnel (that is, in purely physical, *tangible* terms). It should be clear from Buchanan and Huczynski's (1985) definition, however, that the organization is equally an *intangible* reality at the level of both structure and process. Organizational management, in other words, is not simply about the resourcing and allocation of personnel and technological hardware, but is equally, and fundamentally, concerned with providing the necessary *social and psychological environment* which will promote the realization of the organization's goals. With regard to safety and reliability in the workplace, the importance of this *intangible* aspect or character of an organization, and the need to manage it properly, is recognized in the HSE's view that:

> *To prevent accidents to people and damage to plant and the environment one needs to ask how management should be involved. Management's responsibility is to control work – both its human and its physical elements, and accidents are caused by failures of control. They are not, as is so often believed, the result of straightforward failures of technology; social, organization and technical problems interact to produce them.*
> (HSE, 1985.)

The HSE and the CBI (CBI, 1990) see organizational aspects as critical in determining employee behaviour at work and have highlighted the following aspects of management control in their publication (HSE, 1989):

- Clear and evident commitment from the most senior management downwards, which promotes a climate for safety in which management's objectives and the need for appropriate standards are communicated and in which constructive exchange of information at all levels is positively encouraged;
- An analytical and imaginative approach identifying possible routes to human factors failure (this may well require access to specialist advice);
- Procedures and standards for all aspects of critical work and mechanisms for reviewing them;
- Effective monitoring systems to check the implementation of the procedures and standards;
- Incident investigation and the effective use of information drawn from such investigations;
- Adequate and effective supervision with the power to remedy deficiencies when found;
- Effective selection and training which takes account of job and

person specification for tasks (this approach fits the person to the job and should ensure minimum adaptation by the individual to the task).

The principles of safety management will be considered in more detail in Chapter 16. Selection and training are important aspects of organizational control and have particular relevance both for the individual's safe performance and the system's reliability.

Selection and training
Selection and training are both strategies for fitting the person to the job and complement the ergonomic approach discussed in the previous section. The planning of both selection and training should begin with an analysis of the working situation. For training, this means a training-needs analysis, while for selection a job-analysis and subsequent person-specification are required. From these analyses, it should prove possible to design selection and training programmes and set out a framework for their later evaluation and subsequent review (Cox, 1987). In both cases, what is then required is a consideration of:

1. The content and delivery of the processes;
2. The training of those who have to deliver them;
3. The design of the necessary recording and decision making procedures;
4. The marshalling of the necessary resources.

Selection for safety is a contentious issue, and often refers to de-selection, that is, removing individuals from jobs or posts because of their apparent unsuitability as demonstrated, say, by their safety record. In more positive terms, the selection question is posed largely in the context of 'personality' and perhaps accident proneness (or resistance)(Hale and Glendon, 1987).

There may be three strategies to be pursued with respect to selection for safety:

1. Regarding certain key jobs, their should be careful consideration of individuals with a personal or family history of psychiatric illness;
2. The efficient selection of individuals with appropriate knowledge and technical skills and aptitudes for particular tasks;
3. The efficient selection of individuals with appropriate social skills, needs and motivations for working in groups or alone.

Beyond this, it may be sensible to consider selection and training as intimately linked and not separate functions. What cannot be effectively or ethically selected for may be dealt with through training. For example, it might not be possible to select good night shift workers for whatever reason. However, training might provide those employed with strategies for coping with the family and social consequences of shiftwork, for actively managing sleep, and for maintaining vigilance on monotonous tasks during night shifts.

Implications of human factors issues

The importance of human factors considerations is increasingly recognized in the design of safe systems (HSE, 1989). All safe systems of work should not only take into account the limitations of people discussed in earlier sections but also build in safeguards should those limitations be exceeded (see Chapter 5). We have highlighted several examples of human capabilities and fallibilities including information processing limitations, stress and fatigue which could affect task performance, physical limitations of size and reach, individual skills and social and group factors including organizational structure and culture. All these examples relate to the performance-shaping factors discussed in Chapter 5 and highlighted in Table 7.7.

Systems designers should take account of all aspects of human factors in designing and developing safe and reliable systems. Equally, managers and advisers within organizations should select, train and monitor individuals to encourage and sustain safe performance and individuals should take some responsibility for their own safety. A recent model produced by the American Society of Safety Engineers

Table 7.7 Performance shaping factors (PSFs) (after Miller and Swain, 1987)

Internal PSFs

Emotional state	Skill level/previous job history
Intelligence	Social factors
Motivation/attitude	Strength/endurance
Perceptual abilities	Stress level
Physical condition (health)	Task knowledge
Sex differences/age	Training/experience

External PSFs

Inadequate task design
Inadequate workspace and layout
Poor environmental conditions
Inadequate training and job aids
Poor supervision
Unrealistic deadlines

emphasizes the importance of human factors in accident causation (Dejoy, 1990). It relates the antecedent conditions in accidents or systems failures to three factors:

1. The ergonomics of the workstation (including cognitive demands);
2. The management and organizational control;
3. The individual's mental processes (cognitions).

This links in with the three main sections of this chapter and further supports an integrated approach to reliability and safety.

Further reading

Buchanan, D. A. and Huczynski, A. A. (1985) *Organizational Behaviour*, Prentice Hall International, Hemel Hempstead.

Cox, S. and Cox, T. (1996) *Safety, Systems and People*, Butterworth-Heinemann, Oxford.

Eastman Kodak Company (1983) *Ergonomic Design for People at Work*, Vol. 1, S. Rodgers and E. Eggleton (eds), Van Nostrand Reinhold, New York.

Eastman Kodak Company (1983) *Ergonomic Design for People at Work*, Vol. 2, S. Rodgers, E. Eggleton and D. A. Kenworthy (eds), Van Nostrand Reinhold, New York.

Glendon, A. I. and McKenna, E. F. (1995) *Human Safety and Risk Management*, Chapman Hall.

Hale, A. R. and Glendon, A. I. (1987) *Individual Behaviour in the Control of Danger*, Elsevier, Amsterdam.

Norman, D. A. (1988) *The Psychology of Everyday Things*, Basic Books, New York.

References

Best, J. B. (1992) *Cognitive Psychology*, West Publishing Co.

Buchanan, D. A. and Huczynski, A. A. (1985) *Organizational Behaviour*, Prentice Hall International, Hemel Hempstead.

CBI (1990) *Developing a Safety Culture*, Confederation of British Industry.

Cohen, A., Smith, M. J. and Cohen, H. H. (1975) *Safety Program Practices in High Versus Low Accident Rate Companies – An Interim Report*, DHEW Publication No. 75-185, National Institute for Occupational Safety and Health, Cincinnati.

Cox, S. (1985) Women's work: women's health, *Occupational Health*, **37**, 505.

Cox, S. (1987) Safety training: an overview of current needs, *Work and Stress*, **1**, 67.

Cox, T. (1978) *Stress*, Macmillan, London.

Cox, T. (1985) The nature and measurement of stress, *Ergonomics*, **28**, 1155.

Cox, S. and Cox, T. (1996) *Safety, Systems and People*, Butterworth-Heinemann, Oxford.

Cox, T., Leather, P. and Cox, S. (1990) Stress, health and organizations, *Occupational Health Review*, February/March, 13.

Dejoy (1990) Toward a comprehensive human factors model of workplace accident causation, *Professional Safety*, **11**.

Dejoy (1986) A behavioural diagnostic model of self-protective behaviour in the workplace, *Professional Safety*, **26**.

Dodd, D. H. and White, R. M. (1980) *Cognition: Mental Structures and Processes*, Allyn and Bacon.

Eastman Kodak Company (1983) *Ergonomic Design for People at Work*, S. Rodgers, E. Eggleton and D. A. Kenworthy (eds), Van Nostrand Reinhold, New York.

Fitts, P. M. (1951) *Handbook of Experimental Psychology*, chapter on Engineering psychology and equipment design, John Wiley, London.

Hale, A. R. and Glendon, A. I. (1987) *Individual Behaviour in the Control of Danger*, Elsevier, Amsterdam.

HSE (1989) *Human Factors in Industrial Safety*, HMSO, London.

HSE (1987) *Dangerous Maintenance: A Study of Maintenance Accidents in the Chemical Industry and How to Prevent Them*, HMSO, London.

HSE (1985) *Deadly Maintenance: A Study of Fatal Accidents at Work*, HMSO, London.

HSE (1985) *Monitoring safety: An outline report on occupational safety and health by the Accident Prevention Advisory Unit of the Health and Safety Executive*, HMSO, London.

Hollnagel, E. (1993) *Human Reliability Analysis Context and Control*, Academic Press, New York.

Kemeny, J. G. (1979) Report of the President's Commission on the Accident at Three Mile Island, Washington, D.C.

Kroemer, K. H. E. (1989) Engineering anthropometry, *Ergonomics*, **32**, 767.

Miller, G. A. (1956) The magical number seven, plus or minus two: some limits on our capacity for processing information, *Psychological Review*, **63**, 81.

NASA (1978) National Aeronautics and Space Administration, *Anthropometric source book* (Vol. 1), *Anthropometry for designers* (Vol. II), *A handbook of anthropometric data* (Vol. III), Annotated bibliography (NASA Reference Publication 1024, 1978).

Norman, D. A. (1988) *The Psychology of Everyday Things*, Basic Books, New York.

Oborne, D. J. (1995), *Ergonomics at Work*, 3rd ed., John Wiley, Chichester.

Reason, J. T. (1984) Absent mindedness and cognitive control, in *Everyday Memory, Actions and Absent Mindedness,* J. Harris and P. Morris (eds), Academic Press, New York.

Shannon, C. E. (1948) A mathematical theory of communication, *Bell System Technical Journal*, **27**, 379–423, 623–656. Reprinted in *The Mathematical Theory* *of Communication* (1949), C. E. Shannon and W. Weave (eds), University of Illinois Press.

Sheehy, N. P. and Chapman, A. J. (1987) Industrial accidents, in *International Review of Industrial Psychology*, C. L. Cooper and I. T. Robertson (eds) pp. 201–227, John Wiley, Chichester.

Singleton, W. T. (1974) *Man–Machine Systems*, Penguin Books, Harmondsworth.

Chapter 8

Programmable electronic systems

Introduction

Digital computers, based on the thermionic vacuum tube, came into relatively common use as an aid to engineering design in the 1950s. By the mid-1960s transistor-based compact computers such as the Digital Equipment Corporation's PDP8 computer were being used for process control. The development of the integrated circuit, and in particular of large scale integration (LSI), led to the introduction of the microprocessor by Intel in the early 1970s, and then to cheaper and more compact process controllers. These were increasingly used in the late 1970s and 1980s in a wide range of industries.

During the same period computer aided-design (CAD) and the use of computer numerically controlled (CNC) machine tools were introduced. Soon after, the first computer-controlled robots came into operation.

The use of computers in manufacture, process control and, more recently, in high integrity systems is still rapidly increasing. It has brought many advantages. Logic can more easily be built into the system, changes can, in theory, be made more easily and information can be provided to the operator in a more effective form. On the other hand, it has also introduced new safety and reliability problems (see, for example, Neumann, 1985).

These problems are discussed in this chapter. They provide further illustrations of safety and reliability issues and build upon the ideas introduced earlier in the book.

Programmable electronic systems

A programmable electronic system (PES) has at its heart a programmable electronic (PE) unit (see, for example, Figure 8.1). This unit might be a simple microprocessor or it might be a complex computer. The essential property of the unit is that it can be programmed to undertake a range of tasks under software control, that is, under the direct control of the stored programme. The PES hardware consists of the programmable electronic unit, the interfaces through which it inputs and outputs information, the plant actuators with which tasks are performed and the sensors which feed back information. There are also facilities for external control (for example, to initialize the system or to start a sequence of actions) and for storage and display of information.

PES failure modes

Failure modes in PESs have been discussed in a document issued by the UK Health and Safety Executive (HSE) (HSE, 1987a). This document describes two types of failure: random and systematic. Random failures can be expected in the PES hardware. In principle, random failure rates can be predicted (and reduced) using the standard techniques discussed in earlier chapters. However, there are particular problems in the use of these techniques with programmable electronics. For example, there will, in general, be many complex failure modes, and the effects of these failure modes are not necessarily easy to predict. On the other hand, careful design of the system allows many functional checks and diagnostic aids to be incorporated into the software. Redundancy and diversity are employed as discussed in previous examples (see Chapter 3).

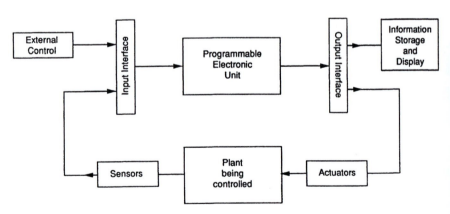

FIGURE 8.1
A programmable electronic system

Systematic failures occur as the result of errors made at the specification, design, construction or operation phase. For example, the equipment must be protected against extremes of temperature, humidity, dust, pollution and mechanical shock. Protection must also be provided against electrical interference and electrostatic effects. Methods of protection are discussed and further references are provided in the second part of the HSE document on PESs (HSE, 1987b). The International Electrotechnical Commission (IEC) in Geneva has also issued three useful documents on the subject (IEC, 1984a–c).

Systematic failures can also result from software errors. Such faults can be very subtle and may remain unrevealed for a considerable length of time until a particular combination of hardware and software conditions is present. Their effect on PES operation will frequently be very difficult to predict.

Random and systematic failures are minimized by the introduction of the highest management and technical standards at all stages of a project. Guidelines were first introduced for military contracts in the United States (MIL-S, 1979; MIL-HDBK, 1981) and the United Kingdom (MOD, 1983). Many other guidelines and codes of practice have been issued since, both nationally and internationally. Current attention is centred on a comprehensive new draft IEC standard entitled 'Functional safety: safety-related systems', IEC (1996).

PES hardware

PES devices are now in common use for a variety of applications. These range from simple microprocessors used to control the suspension or the anti-lock devices on the brakes of automobiles to the fly-by-wire systems which are used to control civil aircraft.

The complexity of the PES will depend not only on the standards of safety and reliability required but also on the complexity of the control or shutdown functions it is called upon to undertake, and on the degree of self-monitoring and testing to be incorporated. The ability to check its own status and performance is one of the particular strengths of the microprocessor-based system. This ability provides a degree of fault tolerance in the PES. When an error is detected steps are taken to apply a correction, to repeat the faulty computation using alternative software or to stop the system in a safe state while remedial action is taken. Higgs (1983) has described a software-based controller for a turbine system that provides a good example of the use of these techniques.

For many tasks a single PES-based controller provides adequate reliability. Reliability can be improved where necessary by employing two processors in parallel sharing the same input and output interfaces. One of the processors now undertakes the control function, its performance being monitored by the second processor. If this detects a malfunction it will give indication of a fault and will take over control.

Since this configuration makes use of shared input and output interfaces it is susceptible to common mode failure in either interface. A higher level of reliability can be achieved by providing separate interfaces to the two processors. Still further enhancement follows from the use of triplicated processors with two interfaces or from three processors and three interfaces. Such highly redundant systems are in common use in a control and monitoring role in the offshore oil industry and in the chemical and petrochemical industries. To obtain the very highest reliability levels with added protection against common mode failures, diversity is introduced. Four-fold redundancy using different microprocessor types, programmed by separate software teams, is commonly used in aircraft flight control systems.

High-reliability PESs are increasingly being constructed using microprocessors specifically developed for the purpose. These are designed to a rigorous specification and have reliability-enhancing features. For example, the VIPER microprocessor (Cullyer, 1987) uses only fixed-point rather than the more common floating-point arithmetic. Programme interrupts and software overlays are avoided and the software has been rigorously tested using formal mathematical procedures.

PES software – software engineering

Software engineering is the name given to the formal and structured approach that is adopted in the production of high quality software. It makes use of the following steps:

1. A definition of requirements is drawn up;
2. A formal specification is produced in such a form that objective checks of the software performance can be made at frequent intervals at the testing and verification stage;
3. A design is produced using modularity as a means of subdividing the programme to ease verification and to minimize inter-modular dependence;
4. Implementation under conditions of close quality control;
5. Testing and verification.

The specification stage is of particular importance. Specification errors have been shown to be a serious source of software safety problems (Griggs, 1981). The specification must define what the system must not do as well as what it must do. At the design stage it is essential to ensure an appropriate allocation of tasks between the human operator and the computer. Some of the problems associated with the human–computer interface are discussed in Chapter 7.

The purpose of the software engineering approach is to produce high quality software with a minimum of faults in it. Two techniques are in common use which can provide reassurance that high reliability standards are being maintained.

First, software reliability models can be used to predict the number of errors remaining in a programme using the observed distribution in time of the errors already found. For such a procedure to be valid, the software development must have taken place under carefully controlled conditions. Software reliability models have been discussed by Bishop and Pullen (1988).

A second completely different approach is taken in static analysis. Typical static analysis packages are SPADE (Carre *et al.*, 1986) and MALPAS (Malpas, 1989). In these, mathematical logic is used to check the logical configuration, data flow and the mathematical expressions in a programme or sub-programme. The results are then compared with those of the programme specification. The technique is described as 'static' in contrast to the dynamic approach involving practical testing of the software used, for example, with reliability modelling. The testing and verification of software is frequently undertaken by a completely different software team to that which developed it. This provides a degree of independence to the procedure.

Having briefly examined the PES and its reliability we are now going to look at two important areas of application of the PES, in robot control and in the computer control of aircraft.

Robots

Robots have been used in industry for well over 20 years. Although the greatest increase in their use has come in the last 10 years or so. Figure 8.2 shows schematically how a robot might be configured to rotate about three horizontal and one vertical axis while its end affector has rotational movement and a gripping action. The various motions are sequenced and controlled by the robot controller unit which is a PES. Power is usually applied hydraulically or electrically.

The simplest type of robot is the pick-and-place robot utilizing end-stops. In this case, the controller initiates a particular movement through the appropriate actuators and this movement continues until end-stop sensors are encountered. Such end-stops are positioned manually and in a point-to-point robot the end-stops are replaced by position measuring sensors. A movement then continues under continuous surveillance of the appropriate position sensors until a pre-programmed position is reached. In a contouring robot, the end effector is made to move along a particular path in passing between two positions. This could be achieved by specifying a large number of consecutive point-to-point moves of very small magnitude. Using the computing power of the controller, however, such consecutive moves can be generated mathematically. Speed of movement can be controlled

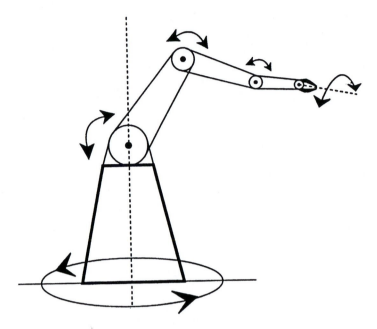

FIGURE 8.2
A robot manipulator

simultaneously. Point-to-point control might be adequate to perform a set of spot welds but contour control is needed for continuous seam welding.

Robot safety

The hazards of operating robotic systems have been discussed on an international scale with a high degree of interchange of information between countries. Barrett (1986) has made a useful inter-comparison of approaches and standards. Robot-related accidents have been analysed by several authors. For example, Carlsson (1985) has investigated 36 such accidents occurring in Sweden between 1976 and 1983. Of these, 14 happened during adjustments in the course of operation and 13 were caused by unexpected movements of the robot during programming or repair. Carlsson's results, which are not atypical, demonstrate that the majority of accidents occur during the relatively brief periods when operators or maintenance staff are working close to the robots.

Programming is undertaken by putting the controller into the 'teach' mode and frequently a teach pendant is used to take the robot, usually at reduced speed, through the movements that it is to repeat in normal operation. It is often necessary to be close to the robot while this happens. The teach pendant normally incorporates a 'dead man's control' which must be continually depressed while the teaching process is under way. Release of the control brings the robot to an immediate halt.

A simple robot installation

Some typical robot safety features are illustrated in Figure 8.3. In this very simple installation a robot is being used for welding operations. The job is located on a work positioner which rotates from the manual loading position to the operating position before welding commences. Both the robot and the work positioner are located in a two metre high opaque enclosure with interlocked access doors. The enclosure prevents unauthorized access while the robot is working. It also provides

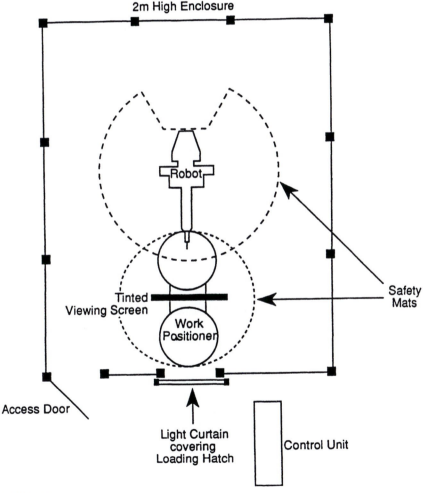

FIGURE 8.3
A robot welding installation

protection against eye damage from the welding flashes and stops those outside the enclosure from being struck by missiles accidentally released by the robot during its operations. The job is loaded through a hatch in the enclosure. A vertical light curtain in front of this hatch ensures that the positioner cannot rotate into the operating position until the operative is clear of the work envelope of the positioner. The robot will not start welding until the positioner has attained the operating position. A tinted viewing screen allows the welding operations to be viewed through the hatch.

Access is needed to the interlocked area for maintenance and for teaching operations. In order to open the interlocked access door, the appropriate key must be withdrawn from the robot control unit. This removes all power from the robot and the work positioner. Insertion of a second key into a teach mode interlock inside the enclosure allows operation of the system under control of the teach pendant. Pressure-sensitive safety mats delineate the working regions of the robot and the work positioner. These are shown dotted in Figure 8.3. They ensure that anyone entering these regions when the teach mode is in use will cause the system to stop under programme control. Emergency 'off' buttons located at strategic points within the enclosure and the 'dead man's control' on the teach pendant all remove power completely. Adequate clearance between the enclosure and the working regions ensures that personnel cannot become entrapped between moving machinery and the enclosure itself.

The safety systems used with robot installations vary greatly from example to example, depending on the hazards involved. In some cases, entry into the enclosure only results in an audible warning. In others it is possible to operate in the teach mode within the working regions of the machines. Protective systems using ultrasound, microwaves, infrared radiation or capacitance effects have been developed for use in such circumstances (see Derby *et al.* (1985), for example).

Robot installations are becoming more numerous and more complex. CAD systems are being increasingly used, not only to determine the layout of an installation but also to undertake an initial programming of the robots off-line. Many new developments can be expected over the next few years.

Computer control of aircraft

Some degree of computer control has been used in military aircraft since the 1970s. In the civil field, Concorde has computer control associated with some functions while the Boeing 757 and 767 and the Airbus A310 and the A300-600 have some degree of computer control associated with the flight control surfaces. The first civil aircraft to have all flight control surfaces operated by means of computers is the Airbus A320. In this aircraft all such control functions pass through the computer system.

The A320 avionics system

The A320 avionics system has been described by Baud (1988). The flight control surfaces are operated by hydraulic systems actuated close to their points of application by digital signals sent by the flight control computers along electronic links, a process known as 'fly-by-wire' (FBW). There are five flight control computers. Three are spoiler and elevator computers (SECs) and two are elevator and aileron computers (ELACs). The SECs and ELACs are manufactured and programmed by separate companies thus providing a high degree of diversity. In addition, the microprocessors incorporated in the two types of computer also have different manufacturers and different programming languages have been used. Each of the five computers is actually two computers in one, one part monitoring the performance of the other. If an error is detected, the relevant control functions are transferred automatically to another computer. The flight control surfaces are physically subdivided with each computer actuating its own sections. There are three independent hydraulic systems to provide the actuation.

The A320 flight control system contains a high degree of redundancy. The aircraft can, in fact, be flown on a single computer. Indeed it could, in the highly unlikely situation of all five computers being inoperable, be flown (and indeed landed) on mechanically operated emergency rudder and tailplane trim controls.

Control of the A320 engines is also exercised by means of a FBW system known as full authority digital engine control (FADEC). The FADEC system ensures that the engines provide the appropriate power for the prevailing flight conditions. The system has the potential to extend engine life and reduce maintenance requirements by optimizing engine running conditions.

The computer systems were thoroughly tested using simulation and verification techniques. Environmental tests included simulated lightning strikes and operation in the presence of intense electromagnetic radiation at radio and radar frequencies.

The computers accept instructions from the pilot through the cockpit controls. Before transmitting the information to the appropriate actuators the computers check that the changes would keep the aircraft within its safe flight envelope. If the pilot attempted to stand the aircraft 'on its tail' or bank excessively steeply, for example, the computers would not comply. Since a great many accidents, in aircraft control as well as elsewhere, are due to human error (Lloyd and Tye, 1982) such surveillance by the computers undoubtedly enhances safety. On the other hand, violent evasive action outside the safe limits set by the computers might, on the rarest occasion, avoid a catastrophic collision, but would be disallowed by the computer system.

Apart from its safety advantages the Airbus 320 avionics system leads to significant weight savings and has clear maintenance advantages both in terms of mechanical simplification and in the way the computer system can minimize component stress and wear.

The Boeing 777 system

This adopts high levels of redundancy and diversity in a similar general approach to that of the Airbus flight controls (Norris, 1994). Flight control laws are used to ensure that flight integrity is maintained, the computers issuing warnings to the pilot in circumstances where transgressions might take place. The pilot maintains the ultimate control authority at all times, however, and can ignore the warnings. This is in direct contrast to the A320 system where the flight control computers have the power of veto over pilot actions.

Future developments

In the near future digital FBW information flow will transfer to fibre optic systems. These will be lighter and less susceptible to outside interference. Another development which could significantly enhance safety is the use of additional auxiliary computers, either on board or on the ground (using a radio link) to monitor aircraft performance and to provide advice to the flight crew when appropriate.

At a more fundamental level, we may see the introduction of unstable aerodynamic configurations into passenger aircraft design. Such configurations can lead to reduced drag on the aircraft and thus to increased fuel efficiency. The 'conventional' aircraft will, in general, continue to fly a steady course if left to fly itself. For an aircraft of unstable configuration this is no longer so. In such a case, positive control must be maintained at all times, thus making a reliable avionics system even more important.

Artificial intelligence

Artificial intelligence (AI) has been defined by Barr and Feigenbaum (1981) as 'the part of computer science concerned with designing intelligent computer systems, that is, systems that exhibit the characteristics we associate with intelligence in human behaviour – understanding language, learning, reasoning, solving problems, and so on'.

AI is being developed for use in many fields such as pattern recognition, language translation, theorem proving and in expert systems (Charniak and McDermott, 1985). In control system applications, AI is starting to find an important role in situations where control decisions have to be made using incomplete or noisy data or where adequate mathematical process models are not available. In this latter case, adaptive 'learning' behaviour can be particularly valuable.

Techniques being applied to these problems include genetic algorithms, fuzzy logic and algorithms based on neural networks.

Failure modes are now even more difficult to predict than in a conventional PES. Amongst the approaches currently being considered to control hazards resulting from such failures is to 'jacket' the AI system within a conventional monitoring system somewhat in the manner of the A320 avionics system's safe flight envelope. Alternatively, the AI techniques can be restricted to subservient roles in which the effects of failure would be strictly limited. Rodd *et al.* (1992) have discussed the use of AI in control systems.

Conclusions

This chapter has considered safety reliability issues in relation to programmable electronic systems. In particular, it has reviewed both hardware and software failure modes and protection systems. It has also considered a number of applications of PESs and their associated safety and reliability.

This chapter marks the first of a series of chapters considering specific system failures. Chapters 9 and 10 consider failures in relation to nuclear and chemical plant.

Further reading

Bonney, M. C. and Yong, Y. F. (1985) *Robot Safety*, Springer Verlag, Berlin.
Charniak, E. and McDermott, D. (1985) *Introduction to Artificial Intelligence*, Addison-Wesley, Wokingham.

References

Barr, A. and Feigenbaum, E. A. (eds) (1981) *The Handbook of Artificial Intelligence*, (Vol. I), Morgan Kaufmann, Los Altos, CA.
Barrett, R. J. (1986) *Robot Safety. Current Thinking and Developments in Major User Countries*, Commission of the European Communities, Health and Safety Directorate, V/E/3.
Baud, P. (1988) The application and integrity of microprocessors in the A320 Airbus, *Electronic Engineering*, July, 39.
Bishop, P. G. and Pullen, F. D. (1988) Probabilistic modelling of software failure characteristics, in *Proc. IFAC Symp. on Safety of Computer Control Systems 1988*, Fulda, Germany, November.
Carlsson, J. (1985) Robot Accidents in Sweden, in *Robot Safety, Current Thinking and Developments in Major User Countries*, Commission of the European Communities, Health and Safety Directorate, V/E/3.
Bonney, M. C. and Yong, Y. F. (eds) (1985) *Robot Safety*, IFS (Publications) Ltd, Springer-Verlag, Berlin.
Carre, B. A. *et al.* (1986) SPADE – The Southampton Program Analysis and Development Environment, in *Software Engineering Environments*, I. Somerville (ed.), Peter Peregrinus, Stevenage.

Charniak, E. and McDermott, D. (1985) *An Introduction to Artificial Intelligence*, Addison-Wesley, Wokingham.

Cullyer, W. J. (1987) Implementing Safety-critical systems: the VIPER microprocessor, *Proc. Workshop on Hardware Verification*, Calgary, Canada, January.

Derby, S., Graham, J. and Meagher, J. (1985) A robot safety and collision avoidance controller, in *Robot Safety*, M. C. Bonney and Y. F. Yong (eds), IFS (Publications) Ltd, Springer-Verlag, Berlin.

Griggs, J. G. (1981) A method of software safety analysis, *Proc. 5th Int. Safety Conference*, Denver, Colorado, System Safety Soc, Vol. I, Part 1, p. III D-1.

Higgs, J. C. (1983) A high integrity software based turbine governing system, *Proc. IFAC SAFECOMP 83*, Pergamon, Oxford, p. 207.

HSE (1987a) *Programmable Electronic Systems in Safety Related Applications, 1. An Introductory Guide*, HMSO, London.

HSE (1987b) *Programmable Electronic Systems in Safety Related Applications, 2*, General Technical Guidelines, HMSO, London.

IEC (1984a) *Electromagnetic Compatibility for Industrial-Process Measurement and Control Equipment; Part 1: General Introduction*, Geneva, Publication 801-1.

IEC (1984b) *Electromagnetic Compatibility for Industrial-Process Measurement and Control Equipment; Part 2: Electrostatic Discharge Requirements*, Geneva, Publication 801-2.

IEC (1984c) *Electromagnetic Compatibility for Industrial Process Measurement and Control Equipment; Part 3: Radiated Electromagnetic Field Requirements*, Geneva, Publication 801-3.

IEC (1996) *Functional Safety: Safety-Related Systems*, Geneva, Publication IEC 1508 (Draft),

Lloyd, E. and Tye, W. (1982) *Systematic Safety*, Civil Aviation Authority, London, p. 7.

MALPAS (1989) *MALPAS user guide*, Issue 4, RTP/4009/UG. RTP Software Ltd, Farnham, Surrey, UK.

MIL-S (1979) *Software quality assurance program requirements*, US Military Specification MIL-S-52779A.

MIL-HDBK (1981), *Evaluation of a Contractors Software Quality Assurance Program*, US Military Handbook MIL-HDBK-334.

MOD (1983) *Guide to the Achievement of Quality in Software*, Ministry of Defence Directorate of Standardization, DEF STAN 00-16/1.

Neumann, P. G. (1985) Some computer-related disasters and other egregious horrors, *ACM Software Eng. Notes*, **10**, 6.

Norris, G. (1994) Genesis of a giant, *Flight International*, 31 August–6 September 1994, 61.

Rodd, M. G., Verbruggen, H. B. and Krijgsman, A. J. (1992) Artificial intelligence in real-time control, *Eng. App. of Artificial Intelligence*, **5**(5), 385.

Chapter 9

Outcomes and consequences

Introduction

Earlier chapters have examined ways by which reliability can be enhanced using such techniques as redundancy and diversity. Despite the use of these techniques, failures still occur. Failure of such high reliability systems usually involves the simultaneous failure of a number of components. The methods used to enumerate these frequently complex failure modes and to calculate their failure probabilities have been outlined. We have also discussed the fundamental importance of human reliability at all stages, including design, commissioning, operation and maintenance. Methods have been described for the enumeration and quantification of failure modes due to human error and for the design of 'user friendly' systems (see Chapter 7).

The next two chapters will consider the outcomes and consequences of failures in high-reliability systems, with particular reference to nuclear and chemical plant. Such plant typically consists of pressure vessels, reactors and distillation columns, interconnected with pipework and controlled by means of valves and pressure controllers. The type of failures under consideration are those in which fluid containment, in the form of piping or a pressure vessel, is breached as happened for example in the Flixborough disaster in 1974 (see Chapter 17). Such events can lead to the uncontrolled release and dispersion of toxic, flammable or radioactive materials.

The purpose of this chapter is to discuss how the dispersion of such materials in the environment can be predicted. We also consider how the effects of the dispersed materials on the exposed population can be estimated. The calculations involved are frequently complex and often contain empirical expressions based on observations of accidental or experimental release of the relevant chemicals. It would be neither practicable or appropriate to attempt a discussion of such calculations. We have limited our approach to a discussion of the main factors influencing release, dispersion and deleterious effects.

Finally, we will discuss the toxicity of chemical substances and aspects of their legal regulation and control.

The source term

Failure of containment leading to uncontrolled release may take place for many reasons. It may, for example, result from the failure of a component or a joint, or the fracture of a pipeline, or from the failure of the containment vessel. The failure of a large pressure vessel can have particularly serious consequences, although thankfully it is a relatively rare occurrence. Such failure may be due to the weakening of the vessel by fatigue, creep or exposure to excessive temperature, or it might result from the over-pressurizing of the vessel due to failure of pressure control mechanisms. In some cases, release can result from forced ventilation following the operation of a pressure-relief valve. This can be necessary when control of pressure is lost or when refrigeration fails in a low-temperature system. Uncontrolled release can also result either from human error (the incorrect valve being opened, for example) or from deliberate human action in the form of sabotage or vandalism (Kletz, 1985). Frequently, HAZOP and other techniques are used to enumerate possible failure modes and their probabilities. Alternatively, a more global approach, based on known failure data from similar plant, can be adapted (see, for example, Davenport, 1983).

The source term predicts, for each failure mode, how much substance is released and how rapidly. For example, the fracture of a pipe might lead to the release of 1 tonne of gas at a rate of 10 kg/s and at a temperature of 5°C. Clearly, the size of the breach will be relevant, as will the nature, temperature, pressure, and capacity of the plant. The source term also takes into account changes that may take place during the release process, such as vaporization or aerosol formation.

Release of gases, vapours or liquids

Substances are released in a variety of phases. The release may be in the form of a gas which is being stored or used under pressure. More commonly, a liquid is released which subsequently vaporizes prior to dispersion. Liquid discharge in the form of leakage is also a common hazard. Bunds (containing walls) are frequently used to control the spread of such leakage. Two phase release of a mixture of vapour and liquid is also encountered. In general, the characteristics of two phase release are more difficult to predict than those of single phase.

Vaporization of liquids

When a release is in the form of a liquid, the rate of vaporization must also be predicted before atmospheric dispersion is considered. The vaporization process will clearly depend on the initial temperature and pressure of the liquid. Three situations are commonly encountered:

1. The liquid is volatile but has approximately the same temperature and pressure as its surroundings. The rate of vaporization will depend on the surface area available. The bund is helpful in limiting this.
2. Superheated liquid, which has been held under pressure is released. Such liquid may start at ambient temperature or at elevated temperature. On release, a certain portion flashes off, that is, it vaporizes rapidly. This process cools down the remaining liquid which then vaporizes somewhat more slowly. A high wind speed greatly increases vaporization rates in these circumstances.
3. A liquid is released at reduced temperature. In such cases the rate of vaporization depends strongly on the rate of transfer of heat from the surroundings, particularly from underneath.

Computer programs have been developed (see Webber and Jones, 1987, for example) which can provide predictions of vaporization in a range of circumstances.

Dispersion

The calculations involved in predicting total quantities released, rates of release and rates of vaporization are not necessarily simple. The results may carry a considerable degree of uncertainty. The precise nature of the containment failure cannot always easily be predicted and the quantity, temperature and pressure of fluid present may vary from time to time during normal operation. Further problems are encountered when atmospheric dispersion is considered. The dispersion of vapour or gas in the atmosphere is influenced strongly by weather conditions.

Weather conditions

Clearly the strength and direction of the wind will be important. Figures are available in the United Kingdom (Page and Lebens, 1986) giving frequencies and strengths of winds as a function of direction. Such figures are normally quoted for measurements at 10 metres above ground level. It is also necessary to know how wind speed varies with height above the ground for various types of terrain (Page and Lebens,

1986). For example, the wind may closely approach a steady maximum at an altitude of 200–300 metres over flat featureless terrain. By contrast, the corresponding altitude over built-up areas containing high-rise buildings may be very much greater. Persistence (length of time before strength or direction changes) and turbulence, are also relevant to dispersion.

Another very important factor is atmospheric stability. Under unstable conditions, temperature decreases relatively rapidly with altitude. For stable conditions, temperature either decreases slowly with altitude or there may be a temperature inversion such that it increases with altitude. The intermediate neutral stability conditions are the most commonly encountered ones in the United Kingdom. These are frequently adopted in dispersion calculations. Stability categories based on these conditions have been defined by Pasquill (1961).

Dispersion with neutral buoyancy

Expressions have been developed (see, for example, Sutton, 1953; Pasquill, 1962; and Gifford, 1961) which can predict the dispersion of gases having a density close to that of the surrounding air. Instantaneous (puff) release and continuous (plume) release are treated separately but, in both cases, a diffusion process providing a normal transverse distribution of concentration is assumed. The dispersion rates depend strongly on atmospheric stability conditions as well as on wind strength. The resulting expressions can be used to predict vertical and horizontal spread of the gas, concentrations at various positions and times and doses integrated over time. Figure 9.1 gives some idea of how strongly dispersion from an elevated source is affected by atmospheric stability conditions.

Dispersion of dense gases

Many hazardous gases are denser than the surrounding air. Gases of high molecular weight (chlorine, for example) have high density at ambient temperature. Others have high density because they are emitted at low temperature. Liquefied natural gas (LNG), for example, is stored at about − 160°C. The main constituent is methane which is buoyant at ambient temperature but LNG will vaporize and disperse as a dense gas at these low temperatures. Dispersal patterns for dense gases differ in a number of ways from those for gases of neutral buoyancy. Dense clouds may spread upwind, because they can 'slump' under gravitational influence. Upward spread is limited and dense clouds with relatively well-defined boundaries tend to be slow to mix with the surrounding air. They can persist at ground level for a long time in the downwind direction and thus frequently present a considerable hazard. Eventually, however, the heavy gas will become

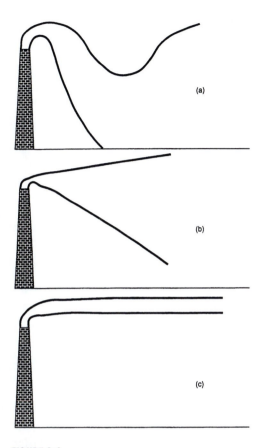

FIGURE 9.1
Dispersion of a gas of neutral buoyancy emitted from an elevated source in (a) unstable,
(b) neutral and (c) stable atmospheric conditions

so diluted by the air that a neutral buoyancy cloud is obtained and
subsequent dispersal will be greatly speeded up. Suokas and Kakko
(1989) have surveyed some recent developments in the prediction of
dense gas dispersion. Valuable experimental data have come from the
Thorney Island and other field trials (McQuaid, 1985).

Other dispersion factors

The dispersion processes so far discussed assume that the gas or vapour
is emitted at low velocity. High-velocity emission is frequently
encountered under conditions of forced ventilation or when pipework
carrying pressurized liquefied gas is breached. Again, gases and vapours
released at high temperature (for example, from a nuclear reactor
accident) can be highly buoyant. In such cases, a rapidly rising plume
will form. The upward movement will reduce as the moving mass cools
down. The resulting contamination can travel many hundreds of miles,

as evidenced by the Chernobyl incident (see Ap Simon and Wilson, 1986, for example). The importance of rain in bringing the material back to earth is also well illustrated by the same incident.

Dispersion can be affected by surface features such as rising or falling ground or the presence of obstructions like trees or buildings. Personnel who are indoors may obtain significant protection against toxic vapour or radioactive clouds. The instruction to remain indoors is given as a part of emergency procedures in Safety Cases (CIMAH, 1984). In order to make realistic predictions of dispersion (and hence of the effects of exposure) all these factors must be modelled.

Consequences of exposure

We have considered briefly the factors influencing the dispersion of gases and vapours. The approach provides predictions of the position and shape of a cloud and of the concentration of gas or vapour at points within it. We wish now to examine the effects of such a cloud on those exposed to it. These will depend on the nature of the material released. Some of the relevant physical and chemical properties were discussed briefly in Chapter 4. Exposure to ionizing radiation is discussed in Chapter 10. The remainder of this chapter considers the effects of fire and explosion, and toxicity.

Fire and explosion

Three conditions must be met if a fire is to start. These are the presence of flammable material, oxygen, and a source of heat to cause ignition. The situation is illustrated symbolically by the fire triangle (see Figure 9.2). Dispersion calculations can predict the region of the gas or vapour cloud in which the concentration is between the lower and upper limits of flammability (see Chapter 4). In this region two of the necessary conditions for fire have been complied with. Only a suitable source of heat is now required.

In practice, many ignition sources may be present, especially when a flammable cloud is located in a relatively large area of human occupation. Possible sources include exposed flames, sparks from welding and cutting, sparks caused by mechanical friction, hot equipment or electrical sparks. The source of electrical sparks can either be electrical equipment or electrostatic discharge.

For the situation where ignition has taken place, an estimate can be made of the resultant thermal radiation intensity (Eisenberg et al., 1975). These authors, using data from military sources, have shown that the probability of death from such thermal radiation will depend on both radiation intensity (I) and exposure time (t) (actually, on $tI^{4/3}$). The expected number of deaths may be predicted using these expressions, when account is taken of the positions of all personnel within the area

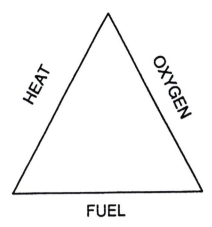

FIGURE 9.2
The fire triangle

affected by the fire. Eisenberg *et al.* (1975) have also published expressions which are valid for pool fires on the surface of flammable liquids as well as for flash fires in gases or vapours, and for non-lethal burn injuries.

Explosion, either by deflagration or detonation can be even more hazardous than fire. Two types of explosion are of particular significance:

1.   The boiling liquid expanding vapour explosion (BLEVE);
2.   The unconfined vapour cloud explosion (UVCE).

A BLEVE can occur when flammable liquid leaks from a storage vessel, ignites and heats up its own or a neighbouring vessel which subsequently explodes. A UVCE takes place when a cloud of flammable vapour or gas having a concentration within the explosive limits is caused to explode by a suitable source of ignition. Destructive effects may be due to the blast wave overpressure, to thermal radiation and, indirectly, to the effects of missiles or by impact against objects.

Eisenberg *et al.* (1975) have given expressions predicting the probability of death due to the blast wave as a function of peak overpressure (p). The destructive power to humans also depends to some extent on rate of rise of pressure and on duration of overpressure. The same authors have provided predictions of non-fatal injuries including eardrum puncture from overpressure and missile and impact damage. In a more general approach, Marshall (1977) has defined the mortality index for explosions as the number of deaths per tonne of explosive material. Figure 9.3, which is based on Marshall's work, plots mortality index against the tonnes of explosive involved. The circular points are mean values for 162 accidental explosions, involving conventional explosives, taken in groups. The square point represents

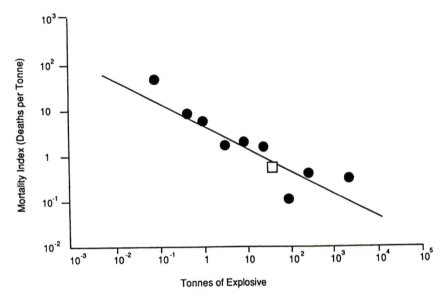

FIGURE 9.3
Mortality index as a function of tonnes of explosive (Marshall, 1987)

the mean of 16 incidents involving fire or explosion following flammable vapour releases. The points suggest clearly that the mortality index decreases with increasing tonnage.

Many authorities have remarked on the high frequency of death and injury resulting from missile impact following explosions. Flying glass is of particular significance. Injury from flying glass can be experienced at relatively large distances from the site of the explosion (see Eisenberg *et al.*, 1975, for example). Fatalities and injuries may also occur from exposure to the chemical release as a result of its toxic properties. These are considered in the next section.

Toxic chemicals

The effects of exposure to toxic chemicals, more recently termed 'substances hazardous to health' (COSHH, 1988), are important in the consideration of both individual and global health risk. The disastrous emission of methyl isocyanate (MIC) from a chemical plant in Bhopal on 2–3 December 1984 (Browning, 1985) heightened public awareness to systems failures in chemical plants and to the concomitant exposures. Lethal MIC gas drifted over the unsuspecting neighbourhood of Bhopal killing at least 2500 people and seriously injuring many more (see Chapter 17). There was a worldwide condemnation and the chemical industry was in the 'dock'. In practice, we are all in daily contact with chemical substances, of varying toxicity, and the developed world is dependent on the chemical and allied industries to sustain it. It is therefore essential that we both exercise adequate safeguards and controls over the location of chemical manufacturing plant (see Chapter 15) and also develop our understanding of chemical substances.

In 1520 Paracelsus, (Boyland, 1982) the father of occupational medicine, wrote: 'All substances are poisons; there is none which is not a poison. The right dose differentiates a poison and a remedy.' Following Paracelsus's principle, it becomes necessary for us to obtain quantitative indices of toxicity to estimate the margin of safety for industrial and other chemicals. It is a function of toxicology to provide this knowledge. Indeed, legal regulation of the safety of industrial chemicals is based on the extrapolation of laboratory toxicity data (see, for example, regulations in the United Kingdom, the United States and other industrialized countries).

Toxicity

The toxicity of a substance is its inherent potential to cause bodily harm and damage to health. This potential will only be realized when the substance comes into contact with or enters the 'victim's' body. The exposure or dose of a given substance will determine the extent of the damage to health. Dose and exposure are often used interchangeably. However, this is not strictly accurate. *Exposure* refers to the amount of substance we may come in contact with, whilst *dose* is the amount which actually enters our body.

For example, we may have an *exposure* of 1 mg of substance per litre of air breathed over a 4-hour exposure period. Our actual *dose* will depend on how frequently we breathe, how much air we breathe each time and the rate at which the body absorbs the toxin.

Substances can gain entry into the body by the following routes:

1. Inhalation through the lungs (the most common route of entry);
2. Ingestion through the mouth;
3. Absorption through the skin;
4. Injection or a wound (less commonly).

All chemicals, however they are absorbed, will find their way into the bloodstream and are then carried to the liver. This organ renders many potentially harmful substances less dangerous by changing their chemical configuration. Occasionally some substances are made more toxic, for example cancer of the bladder arising from beta-naphthyl-amine (Case *et al.*, 1954). The body eliminates harmful substances through the urine, the lungs and less commonly through the skin. Some substances are excreted in the faeces. It is important to understand the biochemical changes and the main excretory routes associated with exposure to toxic substances for meaningful biological monitoring (see Arbetsmiljö, 1984).

A number of factors affect toxicity, some which are substance related and others which relate to the exposed individual or population. In the case of the individual, factors such as age, previous medical history, body weight, gender and lifestyle all have an effect (for example, alcohol and cigarette consumption can affect the health outcome of chemical exposure). Similarly, the toxic effects of substances such as chlorinated hydrocarbons and lead on female reproductive systems and on the developing foetus (teratogenicity) have been the subject of much scientific and political debate. In the United Kingdom such evidence has often culminated in the retention of differential exposure levels for male and female workers (Cox, 1988).

Before any biological experiments or tests are carried out to determine the toxicity of a substance its chemical and physical properties should be considered, together with its chemical structure. If the material is chemically reactive (for example, an alkylating agent) then it is likely to irritate the skin, lungs or eyes. Equally, even when a substance is relatively inert it may still be harmful to humans. A classic example of this may be found in the case of exposure to benzene or asbestos. Some types of toxicity may be predicted from chemical structure and the relationship between chemical structure and biological action are variously discussed by Albert (1978) and Ljublina and Filov (1975). Two important physical properties are physical phase and particle size. For example, water can be more hazardous in its gaseous phase (steam) than as a liquid, and finely divided material is more readily inhaled into the lungs than coarser particles which are filtered out in the nasal passages.

Toxicity tests

Toxicity tests are carried out on animals rather than on humans and fall into three main types:

- Acute toxicity;
- Short-term toxicity;
- Long-term toxicity.

Acute toxicity

The first essential parameter in toxicity evaluation is the acute toxicity, expressed by mortality (death) following administration by appropriate routes. The dose of substance required to cause death is expressed as the lethal dose (LD). If cumulative dose response curves are drawn toxicologists are able to identify doses that affect a given percentage of the exposed group. The commonest is the LD_{50} where 50% of animals will be killed by a particular dose. LD_{50} will often be expressed in terms of mg/kg. This means milligrams of substance per kilogram of body

Table 9.1 Some examples of a classification of toxic compounds (HSE, 1984)

Code	Phrase	Meaning
R28	'Very toxic if swallowed'	LD_{50} less than or equal to 25 mg/kg – acute oral toxicity in rats
R24	'Toxic in contact with skin'	LD_{50} equal to or more than 50 mg/kg but not greater than 400 mg/kg – acute dermal toxicity in rabbit or rat
R20	'Harmful by inhalation'	LC_{50} more than 2 mg/litre/4 hours, but not greater than 20 mg/litre/4 hours – acute toxicity by inhalation in rats

weight of the test animal given by the specified route (for example, by mouth or skin application). Values only give a rough estimate of the degree of toxicity, but this can still be helpful to classify toxic compounds into broad categories. For example they are used as a basis for toxic risk phrases (see Table 9.1) in the United Chemicals (Hazard Information and Packaging for Supply Regulations 1994).

The quantity LC_{50} is also used to express acute toxicity; this refers to the *concentration* which kills 50% of an exposed population (see, for example R20, Table 9.1). LC_{50} will often be expressed in terms of mg/litre/4 hours. This relates to the concentration in mg per litre of air of the substance which the animal breathes for a 4 hour exposure period. It may also be expressed in mg/litre of water/96 hours exposure for fish or aquatic organisms.

The LD_{50} provides a simple measure of toxicity. The full dose–response curve (Figure 9.4) provides a great deal more information, however, and is particularly useful in comparing the effects of two compounds. In our example, the LD_{50} of compound A is less than the LD_{50} of compound B. However, the reverse is true for the LD_{5} for the two compounds.

Short-term toxicity
Short-term toxicity tests follow on from the acute toxicity trials and are used to determine the effects of repeated doses of a substance. Such experiments give an indication of the level which is non-toxic or the 'no effect level'. This is generally accepted to be the level which produces no obvious toxic effect in behaviour or function and does not reduce the rate of growth by more than 10% (Boyland, 1982).

Short-term toxicity tests should also demonstrate whether the material has any cumulative effects and which (if any) organs may be affected by the substance. The usual accepted period for this test is 90 days and it is frequently called the '90-day test'. In reality the tests usually take about six months to complete and experience in this area

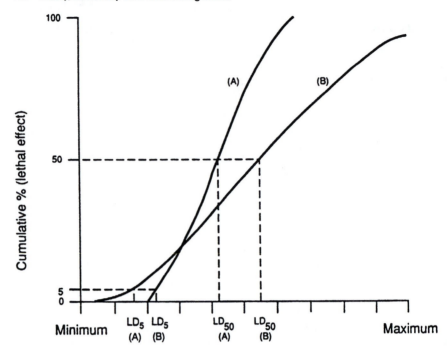

FIGURE 9.4
Full dose–response curves for compounds (A) and (B)

has shown that the majority of adverse effects are seen in the first 30 days (HSC, 1977).

Long-term (chronic) toxicity
One of the main purposes of long-term or chronic toxicity tests is to estimate the carcinogenic (cancer-forming) activity of a substance. However, other pathological effects should not be overlooked and delayed changes which are not neoplastic (growth forming) are known to occur in body tissue. For example, naphthalene can cause cataract of the eye and carbon disulphide may affect the central nervous system (Boyland, 1982). The effects of the testing are seen when the body organs are examined during post-mortem examination.

Chronic toxicity testing is the most expensive of the battery of procedures required to establish the hazards associated with a particular substance and it is therefore advisable to study the data from the 90-day test prior to planning long-term testing. Difficulties in completing safe and meaningful tests include the possibility of experimental animals contracting infections, being attacked or eaten by other animals in the cage and the dangers to laboratory personnel from the potentially hazardous or carcinogenic materials.

Carcinogens

Carcinogens are slow poisons, and it is difficult to know how far the precept of Paracelsus (sometimes known as the first law of toxicology) applies to these, because there may be no safe dose or threshold. In the International Agency for Research on Cancer (IARC) monographs on the carcinogenic effect of chemicals in man it is frequently stated that 'the available data do not allow an evaluation of the carcinogenicity of the compound to be made', thus showing that it is often difficult to determine carcinogenic activity (IARC, 1980).

Cancer risk assessment is essentially a two part process involving:

1. A qualitative judgement of the likelihood of the agent being a human carcinogen;
2. A quantitative judgement of how much cancer the agent is likely to cause at given levels and durations of exposure.

Table 9.2 lists a number of ways in which we can obtain the qualitative and quantitative information on the potential carcinogenicity of chemical compounds. It should be stressed, however, that in the case of the chemical researcher who is synthesizing new chemicals, such evaluations are often impossible. The carcinogenicity of new substances cannot be determined prior to synthesis and may only be assessed by reference to known carcinogens of similar chemical structure. Extreme care needs to be taken by individuals in handling such substances. *In vitro* test methods (Ames *et al.*, 1975) have obvious advantages over the more costly and time consuming animal and epidemiological studies (see later). A detailed account of these methods is given in a publication prepared by the Scientific Group on the methodologies for the Safety Evaluation of Chemicals (SGOMSEC) (Vouk *et al.*, 1985).

Teratogens

An area of increasing concern in recent years has been that of reproductive toxicity. The thalidomide disaster provides us with an example of the effect of toxins on the developing foetus. Substances capable of producing non-heritable birth defects in offspring are called 'teratogens' and thalidomide is a well-known example (Cahen, 1964). Animal studies have demonstrated that the effects induced by a teratogenic substance depend on the degree of foetal development at the time of exposure. In the United Kingdom, teratogens are classi-fied as either Class 1 or Class 2 on the basis of scientific data (HSE, 1989). If there is sufficient evidence to establish a causal relationship between human exposure to a substance and subsequent non-heritable birth defects in offspring, the

Table 9.2 Methods for evaluating carcinogenicity

Method	Advantages	Disadvantages
In vitro, i.e. Ames test, cell transformation test, chromosomal damage evaluation etc.	Simple, fast, cheap; 90% correlation with known carcinogens	Only detects certain carcinogens
Rodent Bioassay	Performed on live animal	Costly, time consuming High doses used Problems in extrapolating to humans
Epidemiological studies	Evaluation in humans possible to estimate risk	Long latent period Costly Difficult to control for all variables
Clinical case reports	Evaluation in humans	Long latent period Statistical evaluation of risk impossible

substance is a Class 1 teratogen. Class 2 teratogens, however, are usually classified on the basis of appropriate animal studies. It is important to understand that teratogenesis involves a disturbance of foetal development and is not concerned with genetic, i.e. heritable damage which occurs in mutagenesis.

Mutagens

These are substances which cause heritable genetic damage. That means they cause changes to genetic makeup which can be passed on to any subsequent children and to their descendants. Mutagens are classified as Class 1 and Class 2 on a similar basis to teratogens.

A battery of tests for mutagenicity has been developed in recent years, following the discovery that many mutagenic chemicals (those having the ability to attack DNA in chromosomes and alter the sequencing of amino acid residues) are also carcinogenic in a specified animal or in man. Since the genetic coding system is so similar for all life forms, chemicals which interfere with DNA of bacteria could also do so in man. Bacterial Mutation tests, such as the Ames test (Ames *et al.*, 1975) are, in fact, the most common of the short-term *in vitro* tests. Unfortunately neither the Ames test nor other *in vitro* short-term tests are totally effective in the detection of human carcinogens. All such tests produce false positives and false negatives.

Potentiation, synergism and antagonism

In most cases the animal and *in vitro* toxicity tests discussed above need to be carried out on a single substance. In real life situations, however, we are exposed to a variety of substances simultaneously. This adds to the complexity of possible toxicological outcomes. Biologically inactive chemicals which enhance the toxic effects of other chemicals are potentiators. Synergism is when two active chemicals produce a combined effect greater than just the addition of the effects of each. Although several examples exist of synergistic action in humans, such as the effect of cigarette smoke on asbestos exposure (Vouk *et al.*, 1985), the importance of synergism at low levels of exposure to carcinogens is not known. Antagonism results when an active chemical reduces the effect of another.

Ecotoxicity

The increasing concern about general environmental effects of chemicals has lead to requirements for testing against ecological effects of their use (see, for example, Vouk *et al.*, 1985). Most commonly, the concern is related to the possible toxic action of the chemical in water courses. The toxicity of a chemical to fish is determined via the LC_{50} test (see earlier). Another important consideration is the biodegradability of the compound and its rate of biodegradation. Toxic chemicals which biodegrade very readily in the aqueous environment will be a limited acute problem, whereas most human disasters have involved persistent chemicals such as organomercury compounds (for example, Minamata Bay, 1973) or polychlorinated hydrocarbons.

Extrapolation of animal toxicity data to humans

The extrapolation of animal toxicity data to humans is a key area of uncertainty in toxicology. In the case of acute toxicity, toxicologists usually work on the assumption that 'humans' are as sensitive as the 'animal'. Table 9.3 gives the equivalent lethal dose based on LD_{50}s in rats.

Table 9.3 Human Equivalents of Various LD_{50}s

Dosage producing death in 50% of treated animals (mg/kg)	Approximate equivalent dose for the average adult human
1	One drop
25	Half a teaspoonful
50	One teaspoonful
100	Half a tablespoonful
200	A tablespoonful
400	Two or three tablespoonfuls
2000	Half a cup

However, if the LD_{50} or LC_{50} test is the sole determinant of acute toxicity in man, erroneous conclusions may be drawn. Early work by various toxicologists showed that for a particular dose–mortality determination the confidence levels depend both on the number of animals and the mortality level (Boyland, 1982). Equally, without an understanding of the underlying mechanisms it is difficult to know how far we may generalize findings across species and situations and how such acute toxicity may respond to interactions between other substances and other activities. This point is even more critical in relation to chronic toxicity, when the very nature of the effects of substances may appear to be quite different. For example, there are major differences between the species in their anatomy and physiology (Boyland, 1982). The respiratory anatomy varies considerably across species and the LC_{50} values of a specific gas will vary accordingly (Withers, 1988). Extrapolation from animals to humans is, in the first instance, also done for young healthy adults. If we are considering the effects of a toxic gas release on the total population we have also to consider the effects on more vulnerable members of the population. Eisenberg and his co-workers (1975) have estimated lethal concentrations of chlorine and ammonia at varying mortalities for all groups of the population. However, even if factors such as vulnerability are taken into account and if data are scientifically collected it is necessary to extrapolate animal toxicity data to human populations with caution.

Epidemiology

Information on the effects of chemicals on humans may also be obtained from epidemiological studies (see, for example, Morison, 1990). Epidemiology is the study of the distribution of disease in relation to populations, in contrast to clinical studies which are carried out at the level of the individual patients. The population at risk is a basic concept in epidemiology. It is important to define this population as precisely as possible. It is also important to consider carefully the way in which this population is sampled if it is too large to be studied in its entirety. The questions of population definition and sampling procedure are critically important in relation to the reliability and validity of the study. Study samples must be derived in such a way as to ensure their representativeness, otherwise findings based on these samples cannot be generalized to the population (Morris, 1983).

Sampling methods

Statistical inference is only possible if the sample is random, or effectively random; that is, each individual in the study population has a known and usually equal probability of selection. Two different techniques are common: first, the simple random sample, and second, the stratified random sample. The former chooses subjects at random from a census or listing of the study population, while the latter, to ensure greater representativeness, first divides the population into subgroups by important variables such as gender or age, and then draws separate random samples from each.

Comparisons in epidemiology

Epidemiological conclusions are based on comparisons. For example, clues to the aetiology of a particular chemical-related/occupational disease may come from comparing disease rates in groups with differing levels of exposure to a suspected chemical. Major studies in this area include those on workers worldwide (Selikoff *et al.*, 1968), results of studies on ischaemic heart disease in a variety of workers (see Morris, 1983) and, more specifically, those carried out in relation to workers in the rubber processing industry on bladder cancer.

Limitations of epidemiological studies

Epidemiological studies are a useful source of data on the aetiology of diseases. However, it is important to remember that such studies can only suggest an association between exposure to a particular chemical and the incidence of disease, they cannot prove causation. Causal proof can only come from experimental studies. However, epidemiological studies can provide a valuable source of data and several different factors can be involved to support the assumption of causality:

1. *Consistency of association* – has the observed relationship been repeatedly observed by different people under a variety of circumstances?
2. *Dose–response relationships* – is the incidence of the disease related in a predictable fashion to the dose of the chemical?
3. *Temporality* – is there a trend with time linking exposure to the chemical with disease?
4. *Plausibility* – does the supposed causal relationship seem biologically reasonable?
5. Can the causal nature of the relationship be demonstrated experimentally?

A recent World Health Organization publication on the epidemiology of accidents and diseases provides a valuable source of epidemiological data from international studies (WHO, 1989).

Legal regulation of toxic chemicals

Legal regulation of safe limits of industrial chemicals, including pesticides, medicines etc., is based (in part) on the assessment of the toxicity. Toxicity data are used to predict the hazards, safe working limits and ultimately safe usage in man. This approach has been developed and is incorporated into the legal requirements in the United Kingdom, the United States, Japan and other industrialized countries. It has also been incorporated into the procedures of the European Community, OECD and, in certain instances, the United Nations.

Frequently, LD_{50} tests are demanded (see above). These use death as the criterion of measurement and thus neglect other biological responses. In the United Kingdom the LD_{50} test has come under increasing pressure under the humanity of animal protection legislation (Animals (Scientific Procedures) Act 1986 and the EEC Directive, 86/609) and has thus been replaced by the Fixed Dose Procedure (Dayan, 1990).

Legal standards are also derived from the evidence of epidemiological studies and, where possible, from industrial records (Levy, 1990). The 'safe' levels of exposure to chemicals in the working environment are expressed as occupational exposure limits.

Occupational exposure: limits and standards
Occupational exposure limits (OELs) have emerged in the western world as a consequence of the work of American industrial hygienists and toxicologists. The US American Conference of Governmental Industrial Hygienists (ACGIH) Threshold Limit Values (TLVs) were accepted and used as a basis for workplace standards. These standards were pragmatic standards and as such they acknowledged factors of control and cost as well as adverse health effects.

In the United Kingdom TLVs gave way to control limits (CLs) and recommended limits (RLs). More recently, Maximum Exposure Limits (MELs) and Occupational Exposure Standards (OESs) have been introduced. MELs are socioeconomic standards, in that, although they are still intended to afford a good level of protection against generally potentially serious health effects which could occur as a consequence of over exposure, they are virtually identical to the previous CLs. They define the maximum permissible loading level. However, employers are expected to reduce concentrations to a level which is as low as reasonably practicable. OESs are health-based standards both by inference and criteria definition. Working to these standards is deemed to be acceptable. Employers are thus required to establish working practices which ensure minimal exposure to toxic substances. In order to assist both the setters and users of standards alike, the United Kingdom Advisory Committee on Toxic Substances (ACTS) has produced and published guidelines (HSE, 1989). MELs and OESs are further explained in EH40 198 (HSE, 1997).

Control of chemical substances is not restricted to control of exposure in the workplace (see, for example, COSHH, 1994). Enforcing authorities worldwide require organizations to control major chemical hazards off-site. In the United Kingdom, this is laid down in the Control of Industrial Major Accident Hazard Regulations, 1984 (CIMAH, 1984). These regulations require organizations to carry out an assessment of the risk of harm from both toxic and thermal hazards in the event of an uncontrolled release. In the case of toxic hazards, the consequences are substance specific, being dependent on its concentration (usually expressed in parts per million (ppm)) and dispersion.

Further reading

Papers in the *Journal of Hazardous Materials.*
Loomis, T. A. (1974) *Essentials of Toxicology,* Henry Kempton.
Sax, N. I. (1989) *Dangerous Properties of Industrial Materials,* Vols. 1, 2 and 3, 7th ed., Van Nostrand Reinhold.

References

Albert, A. (1978) *Selective Toxicity* 6th ed., Chapman and Hall, London.
Ames, B. N., McCann, J. and Yamasaki, E. (1975) Methods for detecting carcinogens and mutogens with the Salmonella/mammalian-microsome mutagenicity test, *Mutat. Res.,* **31,** 347.
Ap Simon, H. and Wilson, J. (1986) Tracking the cloud from Chernobyl, *New Scientist,* 17 July.
Arbetsmiljö, (1984) *Your Body at Work. Human Physiology and the Working Environment,* Distributed by the National Occupational Health and Safety Commission, Box 9, Canberra, A.C.T. 2601.
Boyland, E. (1982) Toxicity testing, in *Occupational Health Practice,* R.S.F. Schilling (ed.), Butterworth-Heinemann, Oxford, p. 563.
Browning, J. B. (1985) After Bhopal, in *The Chemical Industry after Bhopal,* An International Symposium, London, November.
Cahen, R. L. (1964) Evaluation of the teratogenicity of drugs, *Clin. Pharm. Therap.,* **5,** 480.
Case, R. A. M., Hosker, M. E., Drever, B. McD. and Pearson, J. T. (1954) Tumours of the urinary bladder in workmen engaged in the manu-facture of certain dyestuffs and intermediates in the British chemical industry, *Br. J. Industrial Med.,* **11,** 75.
Chemicals (Hazard Information and Packaging for Supply) Regulations, 1994, HMSO, London.
CIMAH (1984) The Control of Industrial Major Accident Hazards Regulations, 1984, HMSO, London.

COSHH (1994) The Control of Substances Hazardous to Health Regulations, 1994, HMSO, London.

Cox, S. (1988) Health and safety: the fender factor, in *The Psychology of Women at Work*, S. M. Oliver (ed.), PSET, Worthing, p. 327.

Davenport, J. A. (1983) A study of vapour cloud incidents – an update, *I. Chem. E. 4th International Symposium on Loss Prevention*, Harrogate, September.

Dayan, A. D. (1990) Chemical hazards and acute toxicity tests, *Biologist*, **37**, 3.

Eisenberg, N. A. *et al.* (1975) *Vulnerability model. A simulation system for assessing damage resulting from marine spills*, Nat. Tech. Information Service Rep. GD-D-137-75, Springfield, VA, USA.

Gifford, F. A. (1961) Use of routine meteorological observations for estimating atmospheric dispersion, *Nucl. Safety* **2**(4), 47.

HSC (1977) *Proposed scheme for the notification of toxic properties of substances*, Health and Safety Commission Discussion document, HMSO, London.

HSE (1997) *Occupational Exposure Limits*, Guidance Note EH40/97, HMSO, London.

IARC (1980) *Long term and short term screening assays for carcinogens. A critical appraisal*, IARC Monographs, Supplement 2, Lyon.

Kletz, T. A. (1985) *What Went Wrong? Case Histories of Process Plant Disasters*, Gulf Publishing, Houston, TX.

Levy, L. S. (1990) The setting of occupational exposure limits in the UK and their interpretation under COSHH, 1988 Regulations. Published in *Occupational Hygiene and Environmental Issues – Control of Chemical Risks on both sides of the factory fence*, Inst. Chem. Eng. Rugby, UK.

Ljublina, E. I. and Filov, V. A. (1975) Chemical structure, physical and chemical properties and biological activity, in *Methods Used in the USSR for Establishing Biologically Safe Levels of Toxic Substances*, WHO, Geneva, p.19.

Marshall, V. C. (1977) How lethal are explosions and toxic escapes?, *Chem. Engr. Lond.*, August, 573.

McQuaid, J. (1985) Objectives and design of the Phase I heavy gas dispersion trials, *J. Haz. Materials*, **11**, 1.

Morison, R. (1990) *Occupational Epidemiology*, 2nd ed., CRC Press.

Morris, J. N. (1983) *Uses of Epidemiology*, 3rd ed., Churchill Livingstone, Edinburgh.

Page, J. and Lebens, R. (1986) *Climate in the United Kingdom*, HMSO, London.

Pasquill, F. (1961) The estimation of the dispersion of windborne materials, *Met. Mag.*, **90**, 33.

Pasquill, F. (1962) *Atmospheric Diffusion*, Van Nostrand, London.

Suokas, J. and Kakko, R. (1989) On the problems and future of safety and risk analysis, *J. Haz. Materials*, **21**, 105.

Sutton, O. G. (1953) *Micrometeorology*, McGraw-Hill, London.

Selikoff, I. J., Hammond, E. C. and Churg, J. (1968) Smoking habits in asbestos workers. Deaths from lung cancer, *J. Am. Med. Assoc.*, **204**, 106.

Vouk, V. B., Butler, G. C., Huel, D. G. and Peakall, D. B. (1985) *Methods for*

Estimating Risk of Chemical Injury: Humans and Non-Human Biota and Ecosystems, John Wiley, Chichester.

Webber, D. M. and Jones, S. J. (1987) A model of spreading Vaporising Pools, *Proc. Int. Conf. on Vapour Cloud Modelling*, Cambridge, Mass., November, J. Woodward (ed.), A. I. Chem. E., p. 226.

WHO (1989) *The Epidemiology of Accidents and Occupational Diseases*, World Health Organization, Geneva.

Withers, J. (1988) *Major Industrial Hazards, Their Appraisal and Control*, Gower Technical Press, Aldershot.

Chapter 10

Ionizing radiation

Introduction

The effects of ionizing radiation are of particular significance for two reasons. First, certain aspects of the subject are of great public concern and must thus feature strongly in public policy making. Second, we know somewhat more about the effects on human health of ionizing radiation than of many toxic chemicals. That this is so only serves to emphasize our relatively poor state of knowledge about both chemical and radiation effects, as will become apparent.

Some basic concepts

The most common types of ionizing radiations are alpha, beta and gamma rays. These have greatly differing powers of penetration through matter, as indicated in Figure 10.1. All three are given off during the process of radioactive decay. Some types of atom have a central nucleus which is in an unstable condition. During the resulting decay process one or more of these types of radiation is emitted. The

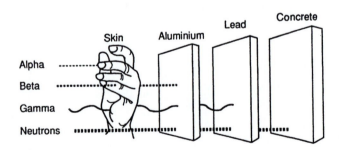

FIGURE 10.1
Schematic illustration of the powers of penetration of various types of ionizing radiation

fourth type of emission illustrated in Figure 10.1 is that of a neutron, which is a constituent part of the atomic nucleus. Some of the very heavy elements such as uranium undergo spontaneous fission, a process during which the nucleus of the atom splits into two parts. When this happens one or more neutrons may also be given off. All four types of radiation, and others, can be produced by particle accelerators. A good example of an accelerator is the x-ray machine producing x-rays for diagnostic or therapeutic use. X-rays are a form of radiation of a similar type to gamma rays.

Effects of ionizing radiation on the human body

Radiation can cause damage to the human body either as an external source or internally, following the intake of radioactive material, most frequently by inhalation or ingestion. The damage may be somatic, that is, direct damage to the victim, or hereditary, damage which arises in the victim's offspring. Such hereditary effects are transmitted as a result of genetic mutations produced by radiation damage in the reproductive system of the victim, (ICRP, 1977). In examining somatic effects, we can identify early radiation effects due to acute exposure. These lead to mass damage to body cells or even to death. Early effects materialize within days or a few weeks. Late effects which may take many years to materialize include various forms of cancer and also cataract formation in the eye.

One final way of classifying damage involves the introduction of probability. Stochastic effects are those where the probability of occurrence depends on the dose of radiation received. Non-stochastic effects are those where severity rather than probability of occurrence depends on the dose received. There may, in some cases, be a threshold below which no harm is apparent. Examples of stochastic effects are hereditary damage and cancer. Non-stochastic effects include radiation burns and cataract formation. Note the equivalence of terminology in dealing with chemical and radiation effects. Acute, chronic, carcinogenic and teratogenic effects of chemicals have equivalents for radiation. The equivalence does not, however, necessarily imply a similarity between the damage processes.

Measurement of radiation doses

The absorbed dose of ionizing radiation is measured by the total energy it deposits in a certain mass of absorbing material. The common unit of absorbed dose is the gray (Gy). A dose of one gray is received when one joule of radiation energy is absorbed in one kilogram of absorbing medium (human tissue, for example). The definition of absorbed dose is a relatively simple one. In practice, however, there are complications. The observed biological damage caused by one gray of radiation

Table 10.1 Current values of the radiation weighting factor

Radiation type	Radiation weighting factor
X-rays, gamma rays and beta rays	1
Neutrons – energy dependent	5–20
Alpha rays	20

depends on the type of radiation involved. To take this into account and to obtain a measure of actual destructiveness to human tissue we define a quantity known as dose equivalent. Thus:

dose equivalent = absorbed dose × radiation weighting factor

If the absorbed dose is in grays, the dose equivalent is in sieverts (Sv). This is a rather large unit. Frequently millisieverts (mSv), thousandths of a sievert, are quoted.

Some examples of current radiation weighting are given in Table 10.1, IRCP (1990). The most damaging types of radiation listed are alpha rays and neutrons of intermediate energy which cause 20 times more damage than the same absorbed dose of x-rays, gamma rays and beta rays.

In many cases where irradiation takes place, the radiation is absorbed in particular parts of the body rather than uniformly over the whole body. Weighting factors are used in such cases to calculate the whole body effective dose equivalent. These factors vary from organ to organ to take account of varying sensitivity to radiation. Internal irradiation produces further complications. When a radioactive substance is taken into the body, the resulting radioactive decay causes internal irradiation. Many such radioactive materials tend to concentrate in a particular organ or type of tissue. For example, radioactive calcium or strontium concentrate in the bones, iodine in the thyroid gland. Two processes put limits on the size of the dose received. First, the radioactivity gets weaker and weaker with time as the radioactive decay process proceeds, and second, the body gradually excretes the radioactive material. The timescales for these processes can vary greatly from case to case.

Annual radiation doses to the public

Table 10.2 summarizes the main sources of ionizing radiation 'available' to individual members of the public in the United Kingdom (NRPB, 1993). The largest single contribution comes from the airborne radioactive gases radon and thoron given off by rocks and soil. Natural radioactivity in the ground and buildings comes next, followed by the radioactivity in foods and the effects of cosmic rays entering the atmosphere from outer space. The largest artificial source of radiation is in the medical use of radiation for diagnostic purposes. 'Products' include radioactivity from gas mantles, smoke detectors and luminous

Table 10.2 Average annual radiation doses in the United Kingdom

Natural origin	mSv	Artificial origin	mSv
Airborne radioactivity	1.30	Medical	0.360
Ground and buildings	0.35	Products	<0.001
Food	0.30	Fallout	0.005
Cosmic rays	0.26	Occupational	0.008
		Discharges	<0.001
Total natural	2.21	Total artificial	0.374

watches. 'Fall-out' includes contributions from the testing of nuclear weapons before 1980 and from Chernobyl. This latter contribution reached its peak in the year following the incident (0.035 mSv) and has declined each year since. Occupational exposure averaged over the population gives 0.008 mSv, while the contribution from the discharge of all types of radioactive waste is less than 0.001 mSv per year. The figures quoted are average ones. There are considerable variations in doses received from person to person.

Dosimetry for ionizing radiation

The dosimetry system for ionizing radiation is based to a great extent on studies of the victims of the Hiroshima and Nagasaki atomic bomb attacks at the end of the Second World War. Within a very short time of the dropping of the bombs Japanese and later American research teams had started gathering information. This work has continued to the present day. By the 1950s the studies were indicating excess numbers of cases of leukaemia amongst the irradiated population (see UNSCEAR, 1977, for example). Other forms of cancer were found to be much slower to develop. As a result, information on these is still coming in.

The situation is illustrated schematically in Figure 10.2. The shapes of the later parts of these curves, particularly curve B, are not at present well established. Provisional dose estimates of the hundreds of thousands of survivors were first issued in 1957 and these were revised in 1965. The 1965 values were used for risk assessment in subsequent years.

In the 1970s an increasing discrepancy was noted between the results from the two Japanese cities. A major re-assessment followed, both of the estimated emission of neutrons and gamma rays by the nuclear explosions and of the estimated radiation doses received by the individual victims. These estimates included consideration of the degree of shielding afforded to the individual by natural features and buildings as well as their distance from ground zero. Moisture content of the air was also found to be significant and was taken into account. The results to date of this massive re-assessment have been published by the Radiation Effects Research Foundation (RERF, 1987). These show that

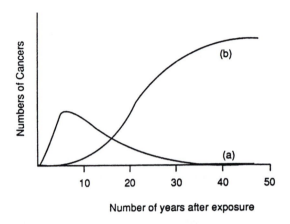

FIGURE 10.2
Times of onset of cancers relative to the time of exposure to ionizing radiation. (a) Leukaemia;
(b) other cancers

there is now an acceptable degree of consistency between Hiroshima and Nagasaki results. A good general account of this work has been given by Bartlett (1988).

The Japanese atomic bomb studies provide information about external irradiation by gamma rays and to a lesser extent by neutrons. Since the irradiation was effectively instantaneous the studies provide no information about dose–rate effects. Experiments on animals (see, for example, Liniecki, 1989) indicate that radiation is less destructive when delivered at lower dose rates. Other information is available from epidemiological studies of workers occupationally exposed to ionizing radiation in the nuclear industry (Doll and Darby, 1987) and in uranium mines (ICRP, 1987). Deliberate medical exposure also provides relevant information. A recent publication by the UK's National Radiological Protection Board (NRPB, 1995), based both on epidemiological investigations and on studies at cellular and molecular level, suggests that there is a finite risk of inducing cancer even at very low doses, i.e. there is no evidence for a threshold effect. The publication confirms previous evidence that risk is greater when a given dose is applied at a high rate than at a low rate by a factor of two to three (see above). The assumption is normally made that the dose–risk relationship remains linear at low doses. The situation is illustrated in Figure 10.3. The linear assumption is that of curve B. The difference between the three curves may not be large at dose D2 but becomes proportionately greater at dose D1.

Early radiation effects have been studied in victims accidentally subjected to acute doses of ionizing radiation. Absorbed doses are normally quoted rather than dose equivalents as quality factors are evaluated for use at relatively low doses. The dose that would be lethal to 50% of those exposed within 30 days, the LD_{50}, is about 3 Gy. Other early effects are listed in Table 10.3.

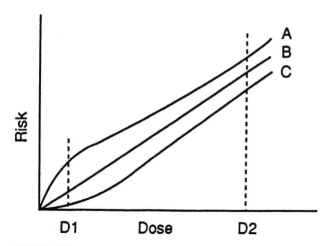

FIGURE 10.3
Possible dose-risk curves

Table 10.3 Early radiation effects in man

Dose (Gy)	Symptoms and effects
1	Nausea and vomiting
1.5	Low risk of death
3	Depletion of white blood cells, death due to infection
10	Gross damage to intestine, death in 3–5 days
50	Damage to central nervous system, rapid death

Risk estimates and exposure limits

Present exposure limits, both for occupational exposure and for the general public are based on recommendations made in 1977 by the International Commission on Radiological Protection (ICRP, 1977).

The recommended limit set by the ICRP for occupational exposure was intended to keep the risk of death due to stochastic effects at a level no greater than the risk of death in other industries having high levels of safety. An annual dose limit of 50 mSv was predicted to lead to a mean occupational dose of around 5 mSv. This was estimated at the time to give a risk of death of 5×10^{-5} per year compared with an industrial figure of about 10^{-4}. The general public was allocated the lower annual dose limit of 5 mSv in the expectation that the average dose to members of the public would thus be held below 1 mSv per year. This latter would lead to a risk of death of 10^{-5} per year, a risk that was considered would be 'acceptable' to members of the public (see Chapter 12). The recommended limit was lowered in 1985 to 1 mSv per year (ICRP, 1985). Such a limit is actually below the average expected dose from radiation of natural origin (Table 10.2).

The ICRP exposure limits are not mandatory – they are recommendations. The current legally enforceable limits in the United Kingdom are contained in the Ionizing Radiations Regulations 1985 and are in line with the relevant European Commission (Euratom) Directive (Euratom, 1980).

Table 10.4 shows the current dose limits for occupational exposure and for exposure of the general public. The occupational limits for the United States, issued by the Nuclear Regulatory Commission (NRC, 1979) are also shown. Limits to the general public, issued by the Environmental Protection Agency (EPA) are source-dependent and vary from 5 mSv per year for commercial reactor radiation releases (EPA, 1985a) to 0.25 mSv per year for emissions into the air (EPA, 1985b). The NRC restricts doses to the public from light water reactors to 0.03 mSv per year from liquid effluents and 0.05 mSv per year from gaseous emissions (NRC, 1986).

Future changes to dose limits

Following the re-assessment of the Japanese bomb data (RERF, 1987), the ICRP and the United Nations Scientific Committee on the Effects of Atomic Radiation (UNSCEAR) published new risk estimates (ICRP, 1987a; UNSCEAR, 1988). These are somewhat higher than earlier estimates. The ICRP subsequently produced revised recommendations on exposure levels (ICRP, 1990). Some changes to legal dose limits can be expected to follow. In Europe the EC has issued to member states a Directive (Euratom, 1996) which limits occupational exposure to 100 mSv in any consecutive five years with a maximum of 50 mSv in any one year. Alternatively, a straight 20 mSv in any year may be adopted if preferred. For members of the public, the limit is reduced from 5 mSv to 1 mSv per year. National legislation will follow soon.

Uncontrolled releases

Uncontrolled releases of radioactive material such as that at Chernobyl can lead to large numbers of people receiving significant doses of ionizing radiation (IAEA, 1986). Methods outlined in Chapter 9 allow predictions to be made of radioactive cloud dispersion and of radiation doses received. Personnel in the path of the cloud will receive immediate radiation doses. These doses may be both external

Table 10.4 Annual exposure limits for occupational exposure and for exposure of the general public

	Occupational (mSv)	General Public (mSv)
UK	50	5
USA	50	Various

and internal, the latter by inhalation. Some victims may exhibit early effects but a far greater number are likely to receive smaller radiation doses leading to late effects. Further radiation doses may follow as the dispersed radioactive material enters the food chain through a complex series of inter-related processes. In these circumstances significant contamination levels may persist for a number of years as was experienced with upland sheep in England, Scotland and Wales following the Chernobyl incident. The situation is clearly complex. Considerable progress has been made in the understanding of the effects of ionizing radiation on humans, but much remains to be done. This is emphasized by the recent reports of excess numbers of childhood leukaemias near nuclear sites in the United Kingdom. Gardner (1989) has shown from epidemiological studies that child-hood leukaemias are more likely if the father is a radiation worker receiving significant doses of radiation. However, there is no evidence at this stage that radiation is the actual cause of the leukaemias. Other possibilities are under consideration, including chemicals or perhaps an infectious agent following population mixing. Inskip (1993) has provided a good summary of the present position.

Further reading

Martin, A. and Harbison, S. A. (1979) *An Introduction to Radiation Protection*, Science Paperbacks 2nd ed., Chapman and Hall, London.
Also papers in *Health Physics* and the *Radiological Protection Bulletin*.

References

Bartlett, D. T. (1988) DS86: New dosimetry for Hiroshima and Nagasaki, *Rad. Prot. Bul.*, **91**, 10.
Doll, R. and Darby, S. C. (1987) Occupational epidemiology: problems of reaching an overview, in *Proc. Conf. Health Effects of Low Dose Ionizing Radiation – recent advances and their limitations*, Brit. Energy Soc., London, p. 105.
EPA (1985a) *National Standards for Hazardous Air Pollutants: Standards for Radionuclides, Final Rules*, Fed. Register 50: 5190.
EPA (1985b) *National Emission Standards for Hazardous Air Pollutants*, Fed. Register 50: 9194.
Euratom (1980), Council Directive 15 July 1980, *Official J. Env. Com.*, No. L 246.
Euratom (1996), *Euratom Directive* 96/29/Euratom.
Gardner, M. J. (1989) Review of reported increases in childhood cancer rates in the vicinity of nuclear installations in the UK, *J. Roy. Statist. Soc. (A)*, **152**, 307.

IAEA (1986) *Summary Report of the Post-Accident Review Meeting on the Chernobyl Accident*, International Atomic Agency.

ICRP (1977) Recommendations of the Commission, *Ann. Int. Com. Rad. Prot.*, 1, Publication 26.

ICRP (1985), Statement from the 1985 Paris meeting, *Ann. Int. Com. Rad. Prot.*, 15, No. 3.

ICRP (1987) Lung cancer risk for indoor exposures to Radon daughters, Ann. Int. Com. Rad. Prot., 17, Publication 50.

ICRP (1987a) Statement from the 1987 Como meeting, *Rad. Prot. Bulletin* 85 (supplement).

ICRP (1990) Recommendation of ICRP, *Ann. Int. Com. Rad. Prot.*, 21, Nos. 1-3 Publication 60.

Inskip, H. (1994) The Gardner hypothesis, *Br. Med. J.*, **307**, 1155.

Liniecki, J. (1989) 'Cancer risk coefficients for high doses and dose rates and extrapolation to the low dose domain', in *The Proceedings of Meeting on the Effects of Small Doses of Radiation*, IBC Technical Services Ltd, London.

NRC (1979) *Radiation Dose Standards for Individuals in Restricted Areas*, US Government Printing Office, 10 CFR 20.101.

NRC (1986) *Numerical Guides for Design Objectives*, US Government Printing Office, 10 CFR 50 (App. I).

NRPB (1993) *Radiation Exposure of the UK Population – 1993 Review*, (National Radiation Protection Board Report R263).

NRPB (1995), *Risk of radiation-induced cancer at low doses and low dose rates for radiation protection purposes*, Documents of the NRPB, 6(1), HMSO, London.

RERF (1987) *US-Japan Joint Reassessment of Atomic Bomb Radiation Dosimetry in Hiroshima and Nagasaki*, Volume 1, W. C. Roesch (ed.), Radiation Effects Research Foundation, Washington and Hiroshima.

UNSCEAR (1977) *Sources and Effects of Ionizing Radiation*, Report of the United Nations Scientific Committee on the Effects of Atomic Radiation.

UNSCEAR (1988) *Sources, Effects and Risks of Ionizing Radiation*, Report to the General Assembly of the United Nations, UN Scientific Committee on the Effects of Atomic Radiation.

Chapter 11

Harm and risk

Introduction

Chapters 9 and 10 have been concerned with how some of the outcomes and consequences of system failure can be modelled and predicted. This has been achieved by first, looking at the degree of exposure to hazard, second, considering the harm that may result and the likelihood of an event with such consequences happening, and finally, the risk. These two chapters focused on predicting the human effect of exposure to toxic gases, radioactive materials and explosions, with particular regard to immediate or delayed death. There are, of course, many other categories (or types) of harm. In this chapter we first consider these in some detail and link the concepts of harm and risk.

Harm – a review

The Institution of Chemical Engineers (IChemE, 1985) has defined hazard as 'a physical situation with a potential for human injury, damage to property, damage to the environment or some combination of these'. The Royal Society (Royal Society, 1992) provides a similar definition but introduces the word 'harm'. 'Hazard' is described as 'the situation that, in particular circumstances, could lead to harm'. We can thus define 'harm' as 'human injury, damage to property, damage to the environment or some combination of these'. This definition extends the focus of harm beyond immediate or delayed death which was considered in earlier chapters.

Various authors, including Marshall (1987a; b), have collated categories of harm. Table 11.1 contains 10 separate categories. (The authors do not

Table 11.1 Categories of harm arising from specified incident

1.	Deaths	Immediate or delayed
2.	Physical injuries	Disabling and non-disabling
3.	Disease	Immediate or delayed
4.	Mutagenic effects	Short- or long-term
5.	Teratogenic effects	Immediate or delayed
6.	Mental 'injuries'	Short- and long-term
7.	Social trauma	Short- and long-term
8.	Disruption of the community	Short- or long-term
9.	Environmental damage	Short- and long-term
10.	Financial loss	Property damage; business interruption; consequential loss

claim that the list is exhaustive or that the order necessarily reflects severity.) Most of the categories will be seen to have both short-term and long-term effects. The harms in Table 11.1 are not rigidly segregated from one another nor is the boundary between 'short-term' and 'long-term' strictly defined.

It is important to distinguish between acute and chronic events in the analysis of consequences and concomitant harms. Acute events (for example, explosions) are short-lived and usually give rise to immediate harm although mental impairment may occur and may also be delayed. Chronic events (for example, continuous or semi-continuous low level toxic emissions) are usually long-lived and give rise to harm over variable time spans. Occupational diseases and some types of environmental pollution provide examples of such harms.

Table 11.1 also indicates that the harm can be sustained by the environment or an organization as well as by people. Such harms are not mutually exclusive and could all occur as the result of a single event.

We now consider the categories of harm listed in Table 11.1 in more detail.

Death

Immediate death is normally a well-defined condition and in countries such as the United States and the United Kingdom, the vast majority of immediate fatalities become known to the authorities. There are thus no significant problems of under-reporting. Quantification of the associated risk is also straightforward, although the cause and circumstances must be specified with care. Numbers of deaths in a particular time interval or numbers of deaths in a particular time interval for a particular number of individuals exposed to the hazard are normally employed as measures. For example, one might quote fatalities per year in a particular industry or fatalities per year per million employees in a particular industry. In some instances, fatalities can be stated for a given number of employees (usually 1000) in a given occupation over the period of their working life. This period is taken to be 100 000 hours, or a total 100 million hours for the 1000 employees. It is known as 'FAR' (Fatal Accident Rate).

Exposure to some 'hazards' or incidents does not cause death instantaneously. In the case of delayed death, the time delay can sometimes make it very difficult to establish a clear connection between the causative agent and the death. This is particularly so when symptoms are shared with other causal agents. It took some time, for example, to establish the link between smoking and lung cancer against a large background of lung cancers due to other causes.

In quantifying delayed deaths some account can be taken of the time delay by specifying mean loss of life expectancy. This can be illustrated numerically. Thus a particular type of immediate fatal injury might be experienced with equal probability at any time in a working life starting at age 17 years and finishing at age 63 years. The mean age at fatality will be half-way between these ages, that is at 40 years. If the mean life expectancy for a 40-year-old is 35 years, the average victim has died at 40 years but would otherwise have lived to 75 years. The mean loss of life expectancy is thus 35 years.

Exposure to a carcinogen might, for similar reasons, take place on average at age 40 years. If the particular type of fatal cancer involved takes on average 15 years to appear, the victims will then have an average age of 40 + 15 years, i.e. 55 years. If the mean life expectancy for a 55-year-old is 23 years, the mean loss of life expectancy is 23 years compared with 35 years for the immediate fatality. More realistic estimates of loss of life expectancy can be made using known statistical distributions instead of the average ages used in our examples. These are listed in actuarial tables.

Physical injuries

This heading covers a whole spectrum of injury varying both in type and severity. Several scales have been developed to measure harms associated with particular injuries, including the Abbreviated Injury Scale (see Yates, 1990). The Abbreviated Injury Scale is a numerical scale which assesses the severity of an injury. Each injury is categorized further by body section. The scale was designed to facilitate uniform data collection and evaluation throughout the world and it has, since 1976, been utilized by a large number of road accident investigators (Petrucelli et al., 1981).

Severity can be measured in terms of the resulting number of days of work lost or the number of days spent in hospital, although real difficulties are encountered in ranking injuries. It is quite impossible, for example, to rank order an accident involving considerable pain and three days off work relative to a second one with little or no pain and six days off work. Again, there is a problem in rank ordering injuries leading to permanent disability relative to those with none. Permanent disability can vary from stiffness in a joint to total paralysis. Many countries employ a system which grades partial disabilities as a percentage of total disability. Such systems are used to determine awards under industrial injury schemes (Munkman, 1985). Burn injuries are classified as first, second or third degree of severity and compensation is awarded in accordance with the classification.

In the majority of cases, the number of incidents decreases with increasing severity as illustrated by Figure 11.1. This is based on the work of Tye (1975) (Figure 11.1(a)) and the Greater London Council (GLC, 1977) (Figure 11.1(b)) in the United Kingdom and of Bird and Germain (1969) in the United States (Figure 11.1(c)). The relative numbers can be expected to vary considerably from hazard to hazard and from industry to industry, but the general trend will be maintained (see, for example, Davies and Teasdale, 1994).

Information used to compile injury statistics comes from a variety of sources. Company records, trades union or employers organizations, government safety authorities or welfare departments and national safety promotion organizations can all contribute to the data pool. Almost inevitably there is a degree of under-reporting which may change over time and often varies between industries. Reporting requirements and categories are different in different countries. Even within a single country these requirements may be subject to frequent change. In the United Kingdom the Health and Safety Executive

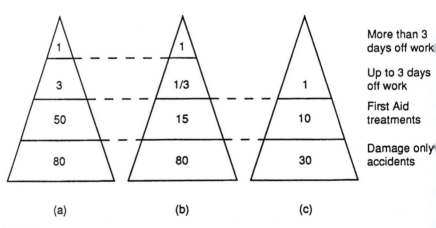

FIGURE 11.1
Accident triangles demonstrating the relative numbers of injuries in different injury categories (see text)

reporting requirements for injuries changed three times between 1980 and 1985.

The risk of injury is normally quoted as the incidence of a particular type of injury per year per 100 000 (10^5) or per million (10^6) employees. These employees may include part-time as well as full-time workers. In some injury databases all employees in a particular workplace are included, whereas others only include the operatives directly at risk. Great care is therefore needed in comparing data from different sources.

Occupational diseases

Occupational diseases vary in severity. In some cases they involve temporary illness with mild symptoms and a limited time off work. However, in other cases they are very much more serious, leading to varying degrees of permanent disability or to death. In line with the system for occupational injuries, partial disability, expressed as a percentage of total disability is used as a measure of severity in some countries. For non-disabling cases, severity can be measured in terms of days off work. Considerable difficulties are encountered when attempts are made to gather information about occupational diseases. Before the data gathering process can even start, there are problems with the recognition of occupational diseases. The many dust-related diseases of the respiratory tract, such as pneumoconiosis in coal miners, byssinosis in textile workers and asbestosis in those handling asbestos, are clearly occupational in nature. So also are the effects of occupational exposure to toxic substances such as lead, cadmium or beryllium. Other well-defined industrial diseases include industrial deafness and occupational asthma (WHO, 1986). For other diseases the attribution of particular cases to specific occupational exposure is difficult. Furthermore, the latency period (the time from exposure for the symptoms to become manifest) is variable. For example, many chemicals and other substances encountered at work are carcinogens (see Chapter 9). Since there can be a delay of many years between exposure to a carcinogen and the onset of cancer it is not always easy to link the cancer with a specific work-related carcinogen. This is particularly so where the type of cancer involved is already common in the community.

Many stress-related disorders have clear origins in the work-place, resulting in a large number of lost working days per year (Cox et al., 1990). Such disorders are not commonly accepted within the United Kingdom. However, the King's Cross fire resulted in damages of £65 000 being awarded to two booking clerks who suffered mental stress and depression after witnessing the tragedy (RoSPA, 1991).

Data on occupation-related diseases are gathered from:

- Returns made by employers to the enforcement authorities;
- Records of welfare authorities responsible for payment of compensation or sickness pensions;
- Death certificates recording identifiable occupational diseases;
- Medical surveillance returns (for example, for blood-lead levels of lead workers);
- Insurance claims;
- Surveys such as census returns or cancer registers;
- Epidemiological studies (see Chapter 9).

It is widely accepted that there is a considerable degree of under-recording of occupation-related diseases, due to failure to identify or to report cases.

Notifiable and prescribed diseases

In the United Kingdom certain diseases, including lead poisoning, toxic anaemia, various lung diseases and decompression sickness, are notifiable by law to the HSE (RIDDOR, 1995). In addition to these notifiable diseases there are a large number of occupational diseases prescribed by the Department of Social Security.

In the United States of America the National Institute of Occupational Safety and Health (NIOSH) has issued guidance on recognition of occupational diseases (NIOSH, 1977).

Mutagenic and teratogenic effects

Mutagens and teratogens have already been discussed in Chapter 9. However, it is important to note that ionizing radiations can produce damage similar to that caused by mutagens, teratogens and carcinogens.

Mental 'injuries' and social traumas

Psychological disorders arising from disasters are becoming increasingly recognized. For example, the incident involving a nuclear reactor at Three Mile Island in 1979 (Kemeny, 1979) illustrates social trauma. At the time of the incident it was headline news throughout the world. There were no immediate fatalities and no physical injuries and the estimate of delayed fatalities due to the radiation effects was two or less. Thus the categories of harm reviewed so far are almost irrelevant in assessing the consequences of this incident.

However, many thousands of local inhabitants were evacuated following the Three Mile Island incident. This caused great disruption to the community. The incident also caused mental injuries on a large scale, and psychological responses varied from:

- Fear and possibly panic;
- Shock to the individual;

- Shock to the community;
- Long term public anxiety;
- Post-traumatic stress disorder (in extreme cases).

Fear, sometimes accompanied by panic, can be expected following other large scale disasters. In such circumstances communications are initially poor so the community has little idea how to act. The emergency services are often stretched to breaking point and it can be many hours or even days before coherent organization becomes apparent. Panic was reported to have followed the liquefied petroleum gas fire at Mexico City in 1984 and the methyl isocyanate release at Bhopal, India, in the same year (see Chapter 17).

Societal shock following large scale incidents can have a strong negative effect on the community. Positive intervention from outside the community can go some way to provide mitigation. This has been discussed in a United Nations report (UNO, 1986). Symptoms of shock are also found in individuals following serious small-scale incidents involving injury.

Long-term public anxiety is likely to be strongest locally but can sometimes be experienced worldwide. The nuclear disasters at Three Mile Island and Chernobyl (IAEA, 1986) led to a significant reduction in public support for nuclear power in many countries. In the United Kingdom it has taken several years for a measure of this support to be regained (Harding, 1990). Public attitudes to risk and acceptability or risk are discussed in Chapters 13 and 15.

Disruption of the community

Disruption to the community following large-scale incidents can occur on a massive scale. Complete communities can be broken up, jobs can be lost and buildings can be destroyed. The most immediate and obvious disruption is that caused by evacuation of the population following an incident. The number of people evacuated provides a measure of the severity of disruption. For example, the release of chlorine at Mississuaga in Canada in 1979 necessitated the evacuation of 240 000 people. At Chernobyl in 1986 120 000 were evacuated and many have not yet been allowed to return.

Environmental damage

Several very large-scale releases of chemicals or radioactive materials can be quoted from recent years where a great deal of damage to the environment has been involved. The Chernobyl incident involved extensive environmental contamination by radioactive material. The crude oil spillage from the ocean tanker *Exxon Valdez* in Alaska in 1989 and the pollution of hundreds of miles of the river Rhine with chemicals discharged following a warehouse fire in Basle in 1986 (LPB, 1987) provide clear examples of large-scale chemical contamination. Some measure of environmental damage can be obtained by evaluating the cost of subsequent remedial work. However, a more detailed account of environmental harms is included within Petts and Eduljee (1994).

One important category remains. Any large-scale incident will have serious repercussions on the company or organization held to be responsible.

Financial loss

There has been an increasing awareness in recent years of just how large the financial losses due to accidents at work may be. Morgan and Davies (1981) and Andreoini (1986) have described costing processes and made estimates (see Chapter 16). More recently, Davies and Teasdale (1994) have made detailed estimates based on interviews at 44 000 households in the United Kingdom. These interviews formed a supplement to the 1990 Labour Force Survey (LFS) and the information was augmented by in-depth studies of accidents at a number of individual companies.

Davies and Teasdale (1994) included the cost of non-injury work accidents as well as those resulting in injury or ill health. They examined the costs to victims and their families, to employers and to society as a whole.

Costs to victims and their families
The following costs were taken into consideration:

1. Short- and long-term loss of earnings;
2. Cost of hospital attendance, medical treatment and other such expenses;
3. Cost of pain, grief and suffering.

The subjective costs (item 3, above) are difficult to estimate. Davies and Teasdale (1994) based their estimates on UK Department of Transport studies of how much members of the public were willing to pay to reduce the risk of death or injury on the roads.

Costs to employers
Some of the costs to the employer are covered by insurance policies, but many others are not. Davies and Teasdale (1994) considered both direct and indirect costs, whether insured or not (Figure 11.2). The costs included were:

1. Payment of compensation for injuries and illness and associated costs. These are usually covered by employer's liability insurance;
2. Loss of output due to absence of injured or ill staff, disruption

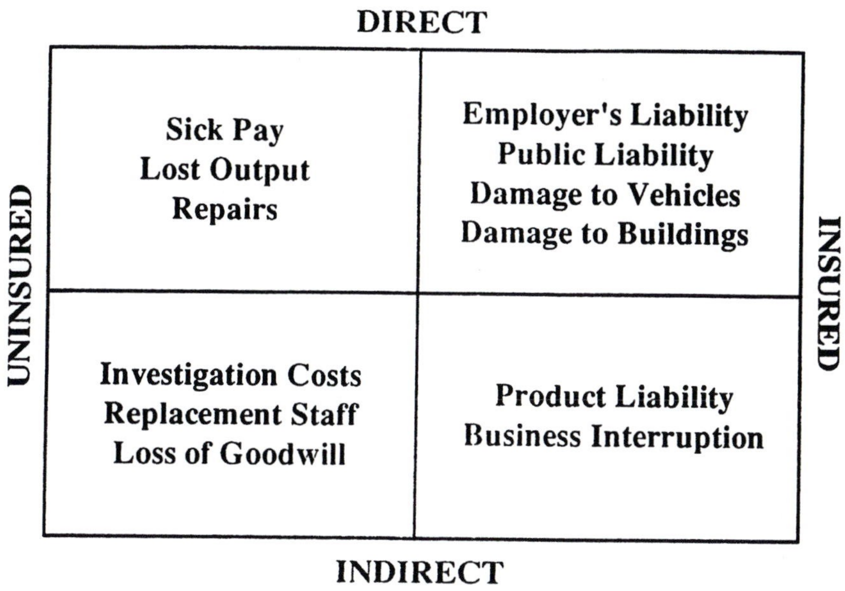
DIRECT
Sick Pay
Lost Output
Repairs
Employer's Liability
Public Liability
Damage to Vehicles
Damage to Buildings
Investigation Costs
Replacement Staff
Loss of Goodwill
Product Liability
Business Interruption
INDIRECT
UNINSURED
INSURED

FIGURE 11.2
Insured, uninsured, direct and indirect costs (Davies and Teasdale, 1994)

of activities and damage to buildings, plant and equipment. Output losses may also result from impaired working ability due to injury or illness while overtime costs may increase as efforts are made to maintain output;

3. Financial penalties for failure to meet contractual deadlines;
4. Cost of hiring and training replacement workers;
5. Cost of accident investigation inducing the disruptive effect on normal activities;
6. Cost of medical treatment provided by the employer;
7. Cost of clearing up and repair. This can induce costs associated with environmental damage;
8. Fines and legal costs;
9. Cost of administration of sick pay;
10. Loss of goodwill and reputation within the workforce and the local community and with customers.

There are problems in estimating the cost of occupational illness, especially in cases where first symptoms may only be apparent many years after exposure (see earlier). Present costs can be calculated but it is not easy to predict the timing of cost reductions which may follow from improvements to present day working procedures. The financial effects of loss of goodwill and reputation (item 10, above) were found particularly difficult to predict and were not included in the final estimates.

Costs to society

In evaluating the costs to society as a whole, care must be taken not to include transfer payments between groups of people. Good examples of such transfer payments are social security payments and the compensation payments made by an employer to the victim. The costs included under this heading were:

1. The cost of loss of current resources such as labour services, materials and capital. The cost of medical treatment is included here. In the absence of accidents these resources would be available for use elsewhere;
2. Losses resulting from infrequent major events such as fires and explosions;
3. The temporary or permanent loss of labour services of victims;
4. The cost of pain, grief and suffering to victims and their families.

Item three presents difficulties when a pool of suitable labour is available due to unemployment. This and many other complications are discussed in Davies and Teasdale (1994).

The study provides a detailed costing for accidents at work. Table 11.2 summarizes overall annual costs to society, to the victims and their families and to employers. To put the numbers into context, the annual cost to society represents two to three per cent of gross domestic product while the cost to employers is equivalent to five to ten per cent of gross trading profits. These are massive costs by any standard. Of particular note in Table 11.2 is the high cost to employers of non-injury accidents. Improved health and safety management standards would undoubtedly lead to a reduction in this cost.

The individual case studies indicate that the ratio of uninsured to insured costs to employers range between 8:1 and 36:1. Insurance only covers a small proportion of costs.

Table 11.2 Annual costs to society, to victims and their families and to employers of workplace accidents

	Workplace injuries £ billions	Work-related illness £ billions	Non-injuring accidents £ billions	Total £ billions
Costs to society	3.49–3.86	4.53–4.72	2.96–7.72	10.98–16.70
Costs to victims and families	1.91	2.72		4.63
Costs to employers	0.85–1.00	0.61–0.74	2.96–7.72	4.43–9.46

Death, injury and industrial disease statistics

The direct measurement of the incidence of various categories of harm can form an important input to policy making in the field of occupational health and safety, for example as an aid to resource allocation in improving control measures. Some relevant statistics are discussed briefly in this section.

Occupational mortality statistics

Occupational mortality statistics are available in the Registrar General's Decennial Supplements. They are often expressed in terms of standardized mortality ratios (SMRs).

The SMR is the ratio of the observed number of deaths amongst an occupational group compared with the number which would have been expected had the mortality rates of the population at large been experienced, allowing for age correction, i.e.

$$\text{SMR} = \frac{\text{Observed deaths}}{\text{Expected deaths}} \times 100$$

An SMR greater than 100 indicates that the group under consideration has a greater than expected mortality whilst the converse is true if the SMR is less than 100.

When interpreting SMRs a number of confounding factors must be borne in mind:

- the effect of social class;
- the small numbers of deaths within certain occupations;
- the process of self-selection, the so called 'healthy worker effect';
- an inaccurate representation of occupational history.

Occupational injury rates

Occupational injury rates have generally been on the decline for several decades in a wide range of countries. This is illustrated for fatal injuries in Figure 11.3. The numbers, taken from an International Labour Office publication (ILO, 1995), are (a) for the United States per million hours worked and (b) for the United Kingdom per 1000 persons employed. The large increase for the United Kingdom in 1988 was caused by the Piper Alpha incident as discussed in Chapter 17. Non-fatal injuries have followed the same general trend. For example, the UK fatal and major injury rate declined from 96 to 81 per 100 000 employees between 1987/88 and 1993/94 (HSC, 1995; HSE, 1992). Temporary increases in these rates in the early 1980s have been discussed by Jones and Tait (1989).

There are large variations in injury rates from one class of the

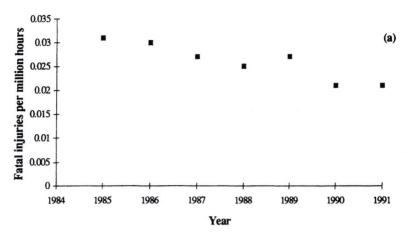

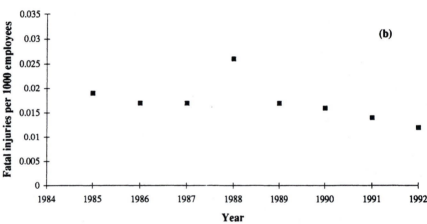

FIGURE 11.3
Fatal injury rates. (a) In the US per million hours worked; and (b) in the UK per thousand employees (ILO, 1995)

Standard Industry Classification (SIC) to another. This can be seen in Table 11.3 (HSC, 1995). In general, injury rates in the service industries SIC6-SIC9 are lower than in the others, particularly so for the more serious injuries. The ratio of serious to all reported injuries varies within the SICs. Much of the variation will be genuine, reflecting differences in the accident triangle (Figure 11.1), but there may also be significant differences in accident reporting efficiencies.

The overall decrease in injury rates discussed earlier is, to a considerable extent, due to an increase in the proportion of the population being employed in the safer service industries.

Table 11.3 Injury rates per 10^6 employees for the various classes of the standard industrial classification (SIC) for the United Kingdom in 1993/94 (provisional)

All reported injuries per 10^6 employees

SIC	0 Agriculture Forestry Fishing	1 Energy Water	2 Minerals Metals Chemicals	3 Metal goods Engineering Vehicles	4 Other Manufacturing	5 Construction
	6950	15478	15118	9042	13034	14163
SIC	6 Distribution	7 Transport	8 Finance	9 Public	All	
	3950	14593	717	5980	7103	–

Major injuries per 10^6 employees

SIC	0 Agriculture Forestry Fishing	1 Energy Water	2 Minerals Metals Chemicals	3 Metal goods Engineering Vehicles	4 Other Manufacturing	5 Construction
	1801	1795	1779	993	1343	2326
SIC	6 Distribution	7 Transport	8 Finance	9 Public	All	
	465	1103	121	604	790	–

Table 11.4 Deaths per year from cancer in various occupations

Occupation	Form of Cancer	Deaths per year per 10^6 at risk
Shoe manufacturing	Nasal	130
Printing	Lung and bronchus	200
Work with cutting oils	Scrotum	400
Wood machining	Nasal	700
Coal carbonizing	Lung	2800
Rubber mill workers	Bladder	6500
Mustard gas manufacturing (1929–1945)	Bronchus	10 400
Cadmium workers	Prostate	14 000
Nickel workers (pre-1925)	Nasal sinuses	6600
	Lung	15 500
Beta-naphthylamine workers	Bladder	24 000

Occupational disease rates

The number of deaths per year from certain occupation-related diseases is still relatively high. For example, there are over 600 deaths from mesothelioma (a form of lung cancer) each year in the United Kingdom amongst asbestos workers (HSC, 1995) compared with about 300 fatal injuries in all industries.

Carcinogens are not always easy to trace, particularly when they induce a form of cancer which is already common in the community. Indeed, Doll and Peto (1981) have suggested that several per cent of all cancers may have occupation-related causes. An ICRP report (ICRP, 1985) lists a number of previously undetected sources of cancer discovered in the last 30 years or so (Table 11.4). The very high mortality rates for many of these compared with the fatal injury rate in Figure 11.3(b) of about 15 per 106 employed should be noted. Once such occupation-related cancers are detected, the causes are normally rapidly eliminated, but deaths can continue for many years due to delays between exposure and appearance of the cancer.

Risk – a review

Risk has been defined (IChemE, 1985) as 'the likelihood of a specified undesired event occurring within a specified period or in specified circumstances. It may be either a frequency (the number of specified events occurring in unit time) or a probability (the probability of a specified event following a prior event), depending on the circumstances'.

The consequence of every 'undesired event' will be harm of some type. Thus, all the categories of harm listed in Table 11.1 may, under appropriate circumstances, have risk associated with them. Most of these categories involve humans but not all. We may, for example, find

Table 11.5 Measures of risk

1. Annual risk (risk of dying in a year)
2. Lifetime risk (the annual risk multiplied by an expected lifespan = 70 years)
3. Risk of specified years of life lost
4. Relative risk (the risk in an exposed group versus an unexposed group)
5. The population-attributable risk (the proportion of deaths in a population or group due to some cause, i.e. occupational exposure)
6. The standardized mortality ratio (the number of deaths in a population expressed as a percentage of the number of deaths in that group if age/sex distribution of the group was the same as the standard population)
7. The margin of safety (the ratio between the highest dose level which does not produce an effect and the anticipated human exposure)
8. Risk of loss of life expectancy (compares the life expectancy associated with an activity with that of a reference set of other activities)
9. Risk of receiving a dangerous dose (HSE criteria)
10. Fatal accident rate (number of fatal accidents suffered by 1000 workers in a particular job or industry over a lifetime)

situations where risks of financial losses associated with environmental damage are present (see Royal Society, 1992). Some of the measures of risk in common use are listed in Table 11.5.

In the remainder of this chapter we wish to pay particular attention to human risk. Two general types of human risk can be distinguished– individual and societal.

Individual risk

Individual risk (IChemE, 1985) is 'The frequency at which an individual may be expected to sustain a given level of harm from the realization of the specified hazards'.

Categories of harm have already been discussed. Thus we might be concerned with the number of delayed deaths per year per million employees leading to ten or more years loss of life expectancy. The level of harm is 'delayed death leading to ten or more years loss of life expectancy'. The frequency of occurrence is the number of such deaths per year per million employees.

Individual risk is experienced by an individual in a clearly defined situation. For example, an assessment of individual occupational risk could be made for a plant operator in a particular work situation while being exposed to known hazards for known (and accurately recorded) working hours. The individual risk in such a case will clearly represent an average value. Risk will change from time to time as plant operating conditions change. It also depends on the distance of the operator from each item of hazardous plant, on wind direction and on the presence of mitigating features and will thus vary as the operators move around performing their duties.

The predicted variation in individual risk with distance around major hazard installations can be plotted as contours of equal risk. Information in this form is commonly used in risk-assessment presentations (see, for example, HSE, 1989).

Societal risk

The second broad category of risk which relates to humans is societal risk. Societal risk is defined (IChemE, 1985) as 'the relationship between frequency and the number of people suffering from a specified level of harm in a given population from the realization of specified hazards'.

The concept of societal risk is useful, for example, in a situation where a large supermarket is being considered for construction close to hazardous plant. Apart from the supermarket staff for whom individual risk could be estimated, the supermarket would be used by a large number of members of the general public, each for a relatively small proportion of their time. We thus have the situation where a significant risk is shared between a large number of people. To merely quote the very small average risk to which the individual was subjected in these circumstances would be highly misleading. An estimate of societal risk provides valuable extra information in such circumstances.

The FN curve

Societal risk is normally expressed in terms of the FN curve. This is best illustrated by means of an example. Chemical plant that produces a particular toxic gas could fail in several different ways, each of these failure modes would lead to release of the toxic gas. In general, the different failure modes would release different quantities of the gas at different rates. For each of these releases it is possible to predict how it would spread in the locality and what the concentration would be at different places as the toxic cloud dispersed. The number of injuries can then be estimated once the locations of people in the vicinity are known. This number will depend on the failure mode and on the positions and numbers of the people, which may vary, for example with the time of day or night.

The results of the predictions are frequently plotted as an FN curve. In this, the cumulative frequency F for all incidents with N or more deaths is plotted against N. For example, on the FN curve of Figure 11.4(a), the point marked indicates that the frequency with which 1000 or more deaths will occur is predicted to be 10^{-4} per year. (A continuous smooth curve has been drawn. However, in our particular example there would have been a series of separate steps corresponding to different failure modes and distributions of people.) The curve also predicts a frequency of 10^{-3} per year for all fatalities, that is for $N = 1$ or more and a maximum number of deaths of about 3000.

The FN curve thus takes some account of the multiplicity of injuries.

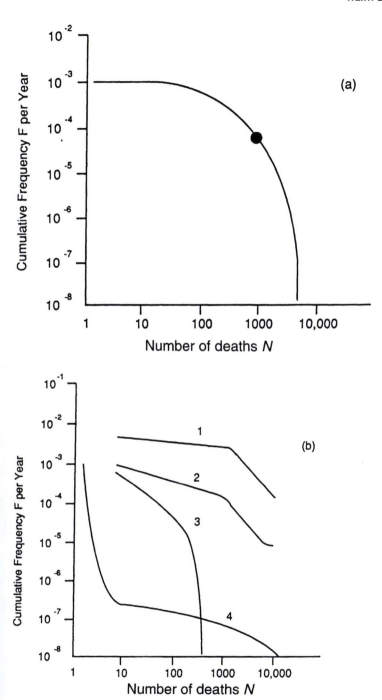

FIGURE 11.4
FN curves. (a) General example; (b) (1) casualties predicted for Canvey before and (2) after improvements, (3) casualties predicted for an explosives wharf and (4) fatalities, including delayed, predicted for the Sizewell B reactor

These may be deaths, non-fatal injuries or any other category of harm. They must be interpreted with great care, as discussed in a publication in the United Kingdom (HSE, 1989).

Figure 11.4(b), based on the HSE publication, shows four *FN* curves. All the curves include serious injuries and fatal injuries apart from (4) which includes fatalities only. Curve (1) is a prediction for Canvey Island on the Thames estuary in the United Kingdom where a proposal had been made for new plant to be constructed at an already busy docking and production complex. The proposal was considered too dangerous until various improvements were made, (Curve (2)), when it was found acceptable and allowed to proceed. Curve (3) is for a proposed wharf for the handling of explosives. Although the curve is well below Curve (2), the development was not considered sufficiently safe to go ahead. Curve (4) was prepared for the enquiry into the siting of a nuclear power reactor at Sizewell in the United Kingdom. Construction of the reactor was permitted following a long planning enquiry.

The *FN* curve can be of considerable aid to the planner and is being used with increasing frequency in that type of application. In practice, there are other factors to be taken into consideration in determining acceptability and these are considered in Chapters 12 and 15. But first we wish to examine the nature of quantitative techniques, their limitations and their uses.

Further reading

Hunter, D. (1978) *Disease of Occupations*, 6th ed, English University Press, London.

Petts, J. I. and Eduljee, G. (1994) *Environmental Assessment for Work Management Facilities*, John Wiley, Chichester.

Ram, W. N. (1983) *Environmental and Occupational Medicine*, Little Brown, Boston, MA.

References

Andreoni, D. (1986) *The Cost of Occupational Accidents and Diseases*, Occupational Safety and Health Series No. 54, ILO, Geneva.

Bird, F. E. and Germain, G. L. (1969) *Damage Control*, American Management Association Inc., New York.

Cox, T., Cox, S. and Leather, P. (1990) Stress, health and organisations, *Occupational Health Review*, February, 11.

Davies, N. V. and Teasdale, P. (1994) *The Costs to the British Economy of Work Accidents and Work-Related Ill Health*, Health and Safety Executive, HMSO, London.

Doll, R. and Peto, R. (1981) The causes of cancer: quantitative estimates of avoidable risks of cancer in United States today, *J. Nat. Cancer Inst.*, **66**, 1191.

GLC (1977) *Cost Value Factors in Accidents and Safety Provisions in the Greater London Council*. Report prepared for the Finance and Establishment Committee of the GLC, London.

Harding, C. (1990) Public understanding: can we make an atom of difference? *Atom*, **401**, March.

HSC (1995) *Health and Safety Commission/Executive Annual Report 1993-94*, HMSO, London.

HSE (1992) Health and safety statistics 1990-91, *The Employment Gazette 100*, September 1992, HMSO, London.

HSE (1989) *Quantified Risk Assessment: Its Input To Decision Making*, HMSO, London.

IAEA (1986) *The Accident at the Chernobyl Nuclear Power Plant and its Consequences*, Vienna.

IChemE (1985) *Nomenclature for Hazard and Risk Assessment in the Process Industries*, Institution of Chemical Engineers, Rugby.

ICRP (1985) *Quantitative Bases for Developing a Unified Index of Harm*, International Commission on Radiological Protection, ICRP Publication 45.

ILO (1995) *Yearbook of Labour Statistics*, 54th Issue, International Labour Office.

Jones, T. S. and Tait, N. R. S. (1989) Occupational health and safety in a changing environment, *Occ. Safety and Health* **19**(9), 10 and 11 i.

Kemeny, J. G. (1979) *Report of the President's Commission on the accident at Three Mile Island*, Washington, DC.

LPB (1987) The Sandoz warehouse fire, *Loss Prevention Bulletin*, **75**, Institution of Chemical Engineers, Rugby.

Marshall, V. C. (1987a) Chernobyl, was it the worst accident? A perspective view on industrial disasters, *Loss Prevention Bulletin*, **81**, Institution of Chemical Engineers, Rugby.

Marshall, V. C. (1987b) *Major Chemical Hazards*, Ellis Horwood, Chichester.

Morgan, P. and Davies, N. (1981) Costs of occupational accidents and diseases in Great Britain, *Employment Gazette*, November, HMSO, London.

Munkman, J. (1985) *Damages For Personal Injuries And Death*, Butterworth-Heinemann, Oxford.

NIOSH (1977) *Occupational Diseases: A Guide to Their Recognition*, Cincinnati.

Petrucelli, E., States, J. D. and Hames, L. N. (1981) The abbreviated injury scale: evolution, usage and future adaptability, *Accident Analysis and Prevention*, **13**, 29.

Ram, W.N. (1983), *Environmental and Occupational Medicine*, Little Brown.

RIDDOR (1995) *The Reporting of Injuries, Diseases and Dangerous Occurrences Regulations*, HMSO, LLondon.

RoSPA (1991) £65,000 for King's Cross clerks, *The Safety Representative*, March(12).

Royal Society (1992) *Risk Analysis Perception and Management*, Report of a Royal

Society Study Group, The Royal Society, London.
Tye, J. (1975) *Accident Ratio Study 1974–75*, British Safety Council, London.
UNO (1986) *Disaster Prevention And Mitigation*, Volume 12, New York.
WHO (1986) *Early Detection Of Occupational Diseases*, Geneva.
Yates, D.W. (1990) Scoring systems for trauma, *Br. J. Med.*, **301**, 1090.

Chapter 12

Quantitative risk analysis: limitations and uses

Introduction

Earlier chapters have described a number of quantitative techniques which can be used to predict the various failure modes of a system, the likelihood of such failures occurring and the consequences of failure. Each of these separate techniques can be incorporated into a logical process of analysis of risks associated with particular events. This overall process is known as quantified risk analysis (QRA) or probabilistic risk assessment (PRA). A publication from the American Institute of Chemical Engineers (AIChemE, 1989) includes a good description of QRA methodology. There are essentially seven stages in its implementation. These are:

1. *System description*, which is the compilation of all technical and human information needed for the analysis (including reliability data).
2. *Hazard identification*, which is a critical step in quantified risk analysis, a hazard omitted at this stage is a hazard which is not analysed.
3. *Incident enumeration*, which is the identification and tabulation of all incidents (or events) without regard to their importance or to the initiating event.

Stages 2 and 3 may be linked together. For example, chlorine gas is a 'hazard' while its unplanned emission through a faulty valve is an 'incident'. HAZOP and other methods are used in hazard identification and incident enumeration(see Chapter 4).

4. *Incident frequency estimation,* which uses likelihood estimation models for selected incidents and evaluates frequencies.

Fault tree analysis (FTA) and event tree analysis (ETA) (see Chapter 4) and the technique for human error rate prediction (THERP) (see Chapter 5) are typical techniques used at this stage.

5. *Consequence estimation,* which is the methodology used to determine the potential for damage or harm from specific incidents (see Chapter 9).
6. *Evaluation of consequences,* this stage is concerned with the estimation of frequency data for specified consequences. Estimates are based on data abstracted from banks (see Chapter 6) and on various sources of historical data (see Chapter 11).
7. *Risk estimation,* combines the consequences and likelihood of all incident outcomes from all selected incidents to provide a measure of risk (see Chapter 11).

QRA, although a powerful tool, is not without its critics. The use of QRA has been greeted with considerable controversy. Within the 'safety and reliability' practitioner community some results and indeed some techniques have been treated with scepticism. Amongst the general public, quantitative results are often regarded with grave suspicion unless they are communicated effectively and are seen by some as being almost irrelevant. Where events with low probability but serious consequences are involved (low F and high N, see Chapter 11), members of the public tend to concentrate their attention on the high N. Some of the communication difficulties associated with QRA are due to poor presentation. Some are due to the fundamental differences between quantitative risks expressed in terms only of F and N and the much broader qualitative approach made by the individual in the perception of risk, as discussed in Chapter 13.

Whatever their limitations, these quantitative predictions provide the only basic technical risk data that are available to the decision maker. However, they must be interpreted with care and judgement. This is frequently misunderstood by members of the public who expect 'scientific' information to be absolute and unequivocal in its predictions. Each assessment must be judged in context and other relevant factors must be taken into consideration. Under such circumstances, quantitative predictions have been shown to be most useful. This chapter discusses the uncertainties involved in the use of QRA and then describes some of the uses to which it is put.

Uncertainties

There will be uncertainties introduced as a result of the possible lack of completeness of the analysis of failure modes and their effects. The inevitable approximations in the modelling of physical processes such as evaporation and dispersion introduce further uncertainties as do inaccuracies in the values adopted for physical parameters and reliabilities. Similar difficulties are encountered in modelling human reliabilities.

Completeness uncertainties

The techniques which are currently used in hazard and risk analysis, including failure modes and effects analysis (FMEA), fault tree analysis (FTA), event tree analysis (ETA) and THERP have been described in earlier chapters. The simple examples discussed in these chapters, however, do not fully illustrate the complexity of real cases or scenarios. In such circumstances there is a very strong possibility of a lack of completeness in the analyses. Thus initiating events, failure modes or even complete physical processes may be accidentally omitted and/or possible consequences may be missed. In particular, common cause failures or consequent failures may be omitted. Techniques like HAZOP and zonal analysis are used to minimize errors due to lack of completeness.

Modelling uncertainties

There are also problems with modelling. These are particularly apparent when predictions are being made of the consequences of events. For example, if a toxic liquid is released it is necessary to provide mathematical models of the release, of the subsequent evaporation and dispersion of the toxic vapour and of the toxic effects on people in the vicinity. There are considerable difficulties in providing mathematical models of these processes and the results of calculations will contain significant inaccuracies. Modelling is also used at the earlier hazard and risk analysis stage, and here again there can be inaccuracies. There are particular problems associated with the modelling of human actions in the event of an emergency and the effects of external events, such as earthquakes or fires. Canter (1980) has developed theoretical models of human behaviour in fires. These are based on his studies of people who have experienced fires taken in conjunction with similar studies carried out for the US Bureau of Standards. The wide range of possible behaviours in fires, together with the paucity of data associated with such events, are illustrative of the problems in this area.

Parameter value uncertainties

Once models have been developed, numerical values have to be assigned to the various parameters. These will include failure rates and distributions, other event rates, temperatures, pressures, physical properties such as viscosity, specific heat and diffusion coefficients, and atmospheric parameters. There will be large uncertainties in the values of many of these quantities.

This is particularly so for failure rate data. Such data are sensitive to component operating conditions and to operator and maintenance standards. Although very large data banks of reliability information are available (see Chapter 6), it is frequently found that the data required for a particular component are not available. In such cases 'engineering judgement' is needed, that is, the values are arrived at by informed guesswork.

For some components it is almost impossible to gather adequate data. Most high integrity equipment is assembled from components of relatively low reliability which are so configured that a much higher overall reliability is obtained. It is then possible to measure the component reliabilities and, subject to the limitations already discussed, to predict the overall reliability. This procedure cannot be adopted where a single component has to operate to an extremely high standard of reliability.

A good example of such a component is the containment vessel for a nuclear reactor. Failure of such a vessel at pressures that it is designed to withstand must have a very low probability, something like 1×10^{-7} per year. For such a low failure rate and for such large expensive objects it is quite impractical to gather accurate failure data. In the circumstances, containment vessels are manufactured to the highest standards, and structural integrity is assessed using a range of non-destructive techniques (see, for example, Tomkins, 1988). Determination of the effectiveness of the testing techniques allows an estimate to be made of the size and numbers of undetected cracks and flaws remaining when the structural integrity tests are complete. Physical theories are then available to link this information to the probability of failure. The procedure is complex and open to considerable uncertainties.

One particularly important form of parameter uncertainty is often encountered: all parameter values are subject to statistical variability. On many occasions it is not possible to test large numbers of components and mean values will carry considerable statistical uncertainty.

Uncertainty analysis
Techniques are available to assess the uncertainties in final estimated quantities in terms of the various component uncertainties. These techniques, which in themselves have created controversy, have been discussed by Vesely and Rasmusson (1984). Accuracy and consistency

of predictions can also be investigated in other ways. Some methods and results are described in the next section.

Accuracy of predictions

A paper by Taylor (1981) provides good evidence of the uncertainties to be encountered in estimating failure rates. Taylor examined the records for almost 10 years of operation of a large hazardous chemical reactor. He compared the fault rates observed for various safety-related sections of the plant instrumentation with the corresponding calculated values. For many components reasonable agreement was found, but in a significant number of cases there were large discrepancies. Some of the worst are given in Table 12.1.

The largest discrepancy was associated with a particular relief valve, predicted to have a failure rate of 0.001 per year and observed to have a mean failure rate of 1.68 per year. Discrepancies were examined in detail. In the case illustrated in Table 12.1 an unknown failure mode was discovered. Other discrepancies were due to instrumentation faults, incorrect component replacement during maintenance, incorrect calibration and undetected component deterioration. Many ambiguities and inconsistencies in fault reporting were encountered due in part to an incomplete definition of 'failure' for components whose performance was gradually deteriorating.

In a second study Snaith (1981) examined reliability and availability records for 146 items of mechanical, electrical and electronic equipment including nuclear power station and chemical plant sub-systems, valves, pressure systems and electronic and computer control equipment. He found that for 64% of the cases the observed values were within a factor of two above or below the calculated ones and in 93% they were within a factor or four.

A comprehensive review of the very few studies designed to compare the validity of hazard risk analysis (HRA) techniques was carried out

Table 12.1 Comparison of expected and observed fault rates

	Expected fault rate per year	Observed rate per year	Ratio
Loss of start-up nitrogen purge	0.1	0.32	3.2
Oxygen valve fails shut	0.5	0.21	0.42
Relief valve opens	0.001	1.68	1680
Isolation of reactor	0.5	0	0
Recycle compressor stops	0.5	1.37	2.74
Loss of reaction	0.01	1.16	116
Loss of compressor power	1.0	4.0	4.0

by Williams (1985). The overall conclusions of the review were that, if high-risk technologists were looking for a HRA method that is comparable in terms of its predictive accuracy to general reliability assessments, the technique of absolute probability judgements (APJ) is probably the best (Humphreys, 1988). However, if the analyst is looking for scrutability, THERP (see earlier) offers the most comprehensive form of modelling. But, as Williams (1985, p.160) concluded, 'if they are seeking methods which are scrutable, accurate and usable by non-specialists, the short answer is that there is no single method to which they can turn. The developers of human reliability assessment techniques have yet to demonstrate, in any comprehensive fashion, that their methods possess much conceptual, let alone empirical, validity'. One way of testing this validity is to carry out a benchmark exercise.

Benchmark exercises

In benchmark exercises, several teams tackle a carefully defined problem independently and compare their results. Two such exercises are described and both relate to the nuclear industry. First, a human factors reliability benchmark exercise, organized by the Joint Research Centre (Ispra) of the European Commission (Poucet, 1988). Second, a technical exercise carried out by 10 different teams from 17 organizations in nine European countries (Amendola, 1986).

Human factors reliability benchmark exercises
In a 'peer review' of THERP Brune and his co-workers (Brune *et al.*, 1983) asked 29 human factors specialists to carry out human reliability assessments (HRAs) on a whole range of possible performance scenarios in a nuclear power plant. For any single scenario they found a wide variation in the problem solutions and HEP estimates varied by as much as five orders of magnitude on some scenarios.

Fifteen teams from eleven countries applied selected HRA techniques to two case studies:

1. The analysis of routine testing and maintenance with special regard to test-induced failures;
2. The analysis of human actions during an operational transient with regard to the accuracy of operator diagnosis and the effectiveness of corrective actions.

The methods used are described in a UKAEA publication (Humphreys, 1988) and included THERP, SLIM and TESEO. In both cases there was a difference of orders of magnitude in the quantitative results. The main contribution to this lack of agreement was the problem of mapping the complex reality of nuclear power generating systems on to these relatively simple models. The exercise also revealed some dangers in using a large and detailed error database, such as that associated with the THERP technique. There was a marked tendency for analysts to model only those errors that appeared in the database and to ignore others that qualitative analyses have found to be important.

Technical reliability benchmark exercise
The teams were presented with detailed technical specifications for an auxiliary feedwater system for a nuclear reactor. The system had been fully designed and was in production, but not yet in operation, so no operational experience had yet been obtained. The 10 teams were set the task of predicting the failure probability. They used their own computer software packages and worked independently, but they met at intervals. Four predictions were made:

1.   An initial predication using their own fault tree and reliability data;
2.   A second following a meeting to discuss and compare qualitative approaches;
3.   A third using a mutually agreed fault tree with their own reliability data;
4.   A fourth using a mutually agreed fault tree and reliability data.

The results are compared in Figure 12.1. The initial prediction varied amongst the ten team results from a lowest failure probability (L) of 6.0×10^{-4} per year to a highest (U) 2.7×10^{-2} per year, a range U/L of 45. The second prediction shows a reduced range of 36 following qualitative comparisons, while still further reduction to 14 follows adoption of a common fault tree. With the introduction of common data, agreement is almost complete, indicating that the different software packages in use were producing results in good agreement. The final agreed failure probability is 2.1×10^{-3} per year. The uncertainty in this value was estimated independently by each team using uncertainty analysis techniques. The results were in reasonable agreement and are all covered by a U/L range of six.

Comparisons of the U/L rates for the third and fourth predictions suggests that parameter value uncertainties are important. It is clear, however, from the second and third prediction ratios that modelling and possibly completeness uncertainties are significant. The improvement from the first to the second prediction to some extent illustrates the difficulties involved in producing a clear definition of the problem in the first place.

Suokas and Kakko (1989) have reviewed studies of the completeness of various HAZOP and other techniques by comparing predicted results with information from failure records. In some cases only 30–40% of all factors had been identified and included, indicating that

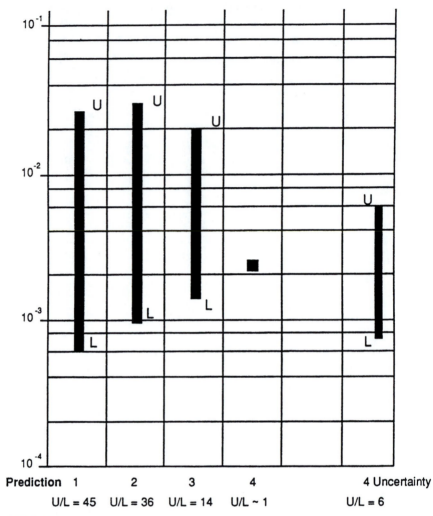

FIGURE 12.1
The results of a benchmark exercise in the prediction of the failure probability of an auxiliary feedwater system

completeness uncertainties may indeed be very significant.

A quantity of less precise information is available to supplement these specifically mathematical studies. Vesely and Rasmusson (1984) have pointed out that relative evaluations of probability or frequency will in general be more accurate than absolute ones. They suggest that unavailabilities can normally be credible to a factor of three, while for accident frequency estimates a factor of 10 is more appropriate. For very low frequencies (1×10^{-9} per year) they suggest that results will only be credible to a factor of 100 in some cases. Daniels and Holden (1983) come to similar conclusions.

Accuracy of consequence predictions

There will certainly be modelling and parameter value uncertainties in consequence predictions and completeness uncertainties may well also be present. Frequently, simplifications are introduced into modelling processes in order to make calculations less complex. Again, many mathematical models contain empirical constants which are adjusted to obtain good results in a particular situation. However, predictions may not be as good in other situations. Model weaknesses have been discussed in a paper by Suokas and Kakko (1989).

Experimental releases of mainly heavy gases have provided direct testing for gas dispersion models. Experiments performed at Maplin Sands, China Lake, Frenchman Flat and Thorney Island have been described and compared by Puttock and Colenbrander (1985). Such tests typically predict gas cloud centroid position, cloud height, cloud horizontal dimensions, maximum concentrations and other such quantities. Results usually agree within a factor of two or three, see papers by Alp (1985), Frayne (1985) and others at the 1985 Toronto Heavy Gas Workshop. Figure 12.2 based on Frayne (1985) shows

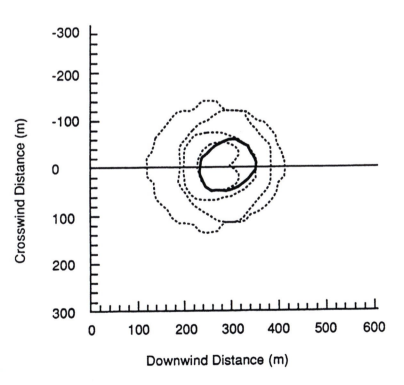

FIGURE 12.2
Release of a heavy gas showing the position and size of the cloud after 90 seconds. Full line – experimental observations; dotted lines – predictions of four theoretical models

experimental data 90 seconds after release from the Thorney Island tests compared with four different predictive models. Downwind drift is well predicted while cloud size maximum discrepancy is about a factor of two. When modelling for source terms, features of the terrain, mitigating features and effects of exposure are added, far greater uncertainties can be expected.

Uses of quantitative techniques

The use of simple probabilistic calculations was introduced extensively in World War II as an aid to strategic and tactical policy making (see Blackett, 1962, for example). The development of reliability and availability prediction followed soon after and is described in Chapters 2 and 3. The use of such quantitative techniques became very much more general, and more controversial, following the introduction in the 1960s and 1970s of fault tree analysis and related techniques.

We have seen in earlier sections that predictions can carry very large uncertainties – a factor of 10 is not uncommon. In such circumstances it is not unreasonable to question whether such predictions are worth attempting at all. The argument that they frequently provide the only quantitative information that is available has already been put forward. However, there are several more positive points to be made.

First, risk practitioners commonly point out that the very use of highly structured techniques such as fault and event tree analysis encourages a careful systematic approach. Such an approach is likely to be valuable in safety and reliability terms, quite independent of numerical results. Substitution of even quite inaccurate values can therefore provide valuable information. For example, it may show that some branches of the tree have so much lower probability than others that they are quite unimportant. A good example of this is to be found in the discussion of Figure 4.6.

Again relative studies, in which one or more values are changed as a way of testing the effect on the final result, can provide valuable engineering insight. In the chemical industry probabilistic techniques were used at first as part of the design process (Hensley, 1968; Stewart, 1971). In nuclear power the main thrust was to use them as an important component of the case to licensing authorities. The first large-scale presentation of this type was the WASH-1400 reactor safety study (Rasmusson, 1975). Similar presentations are now in common use in applications to planning authorities to build hazardous plant at a particular location or to undertake other building developments in the vicinity of existing hazardous plant. Good examples are, respectively, the Sizewell 'B' pressurized water reactor Inquiry and the Canvey Island Inquiry (HSE, 1978) in the United Kingdom. Accident and incident investigations commonly make use of quantitative information as in the Three Mile Island report (Kemeny, 1979) and such information is invaluable in planning emergency procedures and civil defence requirements.

Mention has already been made of the use of techniques such as FTA, ETA and THERP as part of the normal design process. This started in chemical and nuclear engineering and aircraft design but rapidly spread to a range of activities including the design of medical apparatus, power distribution, offshore oil and space research (see Green, 1982, for example) and into environmental policy making (Russel and Gruber, 1987). These techniques are also being used as an aid at the project operation stage (Holloway, 1988). Here they can be continuously updated to account for changes and can be used:

- To evaluate proposed changes;
- To determine safety improvements and allocate priorities;
- To optimize maintenance procedures;
- As an aid to staff training.

QRA techniques as used in the nuclear power industry have been reviewed by Wu and Apostolakis (1992) and in civil aviation (Tait, 1994).

One final application of risk and consequence predictions may be mentioned. This is in risk management, a procedure used to identify, evaluate and control risks. Risk management is an established technique in financial control of high risk prospects and in insurance (see Chapter 16). The use of quantitative techniques in policy making is discussed in detail in Chapter 15.

References

AIChemE (1989) *Guidelines for Chemical Process Quantitative Risk Analysis*, Centre for Chemical Process Safety of the American Institute of Chemical Engineers.

Alp, E. (1985) COBRA: An LNG model, *Proc. Heavy Gas (LNG/LPG) Workshop*, Toronto, Canada, 29–30 January 1985. R. V. Portelli (ed.), Concord Scientific Corporation, p. 76.

Amendola, A. (ed.) (1986) *Systems reliability benchmark exercise, Final Report, Part 1 – description and results*, Commission of the European Communities, Joint Research Centre, Ispra, EUR 10896 EN/1.

Blackett, P. M. S. (1962) *Studies of War*, Oliver and Boyd, Edinburgh.

Brune, R. L., Weinstein, M. and Fitzwater, M. E. (1983) *Peer Review Study of* the *Draft Handbook for Human Reliability Analysis with Emphasis on Nuclear Power Plant Applications*, NUREG/CR-1278, Sandia National Laboratories, Alburquerque, NM.

Canter, D. (1980) *Fires and Human Behaviour*, John Wiley, Chichester.

Daniels, J. T. and Holden, P. L. (1984) Quantification of Risk, *Proc. 4th Int. Symp. on Loss Prevention and Safety Promotion in the Process Industries*, Harrogate, England, 12-16 September 1983. Inst. Chem. Eng. Symposium Series No. 80, p. 33.

Frayne, R. (1985) Session Chairman's introductory remarks, *Proc. Heavy Gas (LNG/LPG) Workshop*, Toronto, Canada, 29–30 January 1985, R. V. Portelli (ed.), Concord Scientific Corporation, p. 108.

Green, A. E. (ed.) (1982) *High Risk Technology*, John Wiley, Chichester.

Hensley, G. (1968) Safety Assessment – a method of determining the performance of alarm and shutdown systems for chemical plants, *Measurement and Control*, **1**, T72.

Holloway, N. J. (1988) Past PSA studies and applications, in *Nuclear Safety after Three Mile Island and Chernobyl*, G. M. Ballard (ed.), Elsevier, Amsterdam, p. 34.

HSE (1978) *Canvey: An investigation of potential hazards from operations in the Canvey Island/Thurrock area*, HMSO, London.

Humphreys, P. (1988) *Human Reliability Assessors Guide*, United Kingdom Atomic Energy Authority, Warrington.

Kemeny, J. G. (1979) (Chairman), *Report of the President's Commission on the accident at Three Mile Island*, Washington, DC.

Puttock, J. S. and Colenbrander, G. W. (1985) Dense gas dispersion – experimental research, *Proc. Heavy Gas (LNG/LPG) Workshop*, Toronto, Canada, 29–30 January 1985, R. T. Portelli (ed.), Concord Scientific Corporation, 32.

Rasmusson, N. (1975) *The reactor safety study: An assessment of accident risks in US commercial nuclear power plants*, Report WASH-1400, NUREG - 75/014. Nuclear Regulatory Commission, Washington, DC.

Russel, M. and Gruber, M. (1987) Risk assessment in environmental policy-making, *Science*, **236**, 286.

Snaith, E. R. (1981) *The correlation between the predicted and the observed reliabilities of components, equipment and systems*, United Kingdom Atomic Energy Authority, National Centre for Systems Reliability report NCSR 18.

Stewart, R. M. (1971) High integrity protective systems, *I. Chem. E. Symposium Series*, 34, 99.

Suokas, J. and Kakko, R. (1989) On the problems and future of safety and risk analysis, *J. Haz. Materials*, **21**, 105.

Tait, N. R. S. (1994) Reliability, safety and civil aviation, *Aeronautical J.*, **98**, 185.

Taylor, A. (1981) Comparison of predicted and actual reliabilities on a chemical plant, *I. Chem. E. Symposium Series*, 66, 105.

Tomkins, B. (1988), Structural integrity assessment methods, *Proc. Seminar on the structural integrity of nuclear reactors*, London, September, Inst. Mech. Engs., London.

Vesely, W. E. and Rasmuson, D. M. (1984) Uncertainties in nuclear probabilistic risk analyses, *Risk Analysis*, **4**, 313.

Wu, J. S. and Apostolakis, G. E. (1992) Experience with probabilistic risk assessment in the nuclear power industry, *J. Haz. Materials*, **29**, 313.

Chapter 13

Risk assessment and cognition: thinking about risk

Introduction

A number of the earlier chapters have discussed the development of methods for calculating the risk from various hazards or potential hazards. These 'risk estimation' measures have been based on, for example, the number of times system failures have occurred or the number of deaths or injuries caused by an activity (mainly occupational) in a designated period of time. Table 13.1 illustrates the levels of fatal risk associated with certain activities expressed as the probability of death ranging from 1 in 1000 per annum for relatively high-risk activities to 1 in 10 million for low-risk events. It introduces the concept of relative risk and contrasts mortality estimates for high-risk groups within relatively risky industries with the corresponding groups in the safest work environments.

Table 13.1 Mortality estimates for various hazardous activities. Level of fatal risk (average figures, approximated)

Per annum	
1 in 1 000	Risk of death in high risk groups within relatively risky industries such as mining
1 in 10 000	General risk of death in a traffic accident
1 in 100 000	Risk of death in an accident at work in the very safest parts of industry
1 in 1 million	General risk of death in a fire or explosion from gas at home
1 in 10 million	Risk of death by lightning

We have also introduced the process of quantified risk analysis as the identification of hazards, the assessment of their mechanisms of harm and the consequences, and of the probabilities with which any and all of these will occur (see Chapter 12). Although such calculations are not always accurate, they have been described 'the best of their kind at the present moment' (Tait, 1995).

Quantified risk analysis (QRA) does not take into account judgements about the significance of hazardous events, etc., and risk levels, as perceived by individuals and by the various groups of individuals which make up our society. This area of concern has been termed 'risk evaluation' (see Figure 13.1). It is part of the overall process of risk assessment (the final stage in QRA) and it also requires the introduction of acceptance standards. However, we need to consider the following issues before acceptance criteria can be established:

1. Individual cognitions (perceptions, knowledge and under-standing) of hazards and risks;
2. The role of the various legislative bodies;
3. The competence and awareness of 'risk assessors';
4. The nature of the risk-decision making process itself.

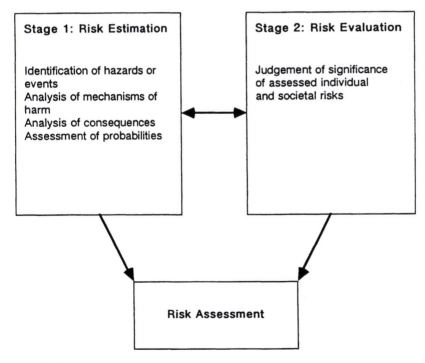

FIGURE 13.1
Quantified risk analysis

This chapter focuses on individual cognitions of hazards and risks. It first discusses the process of risk evaluation and describes a number of related studies, and finally it considers how acceptance criteria are developed as a consequence of these studies.

Risk evaluation

The evaluation of any risk is dependent on the person's (or group's) perception and knowledge of that hazard and the associated consequence and, in turn, their experience of that or similar hazards. The role of cognitive (mental) processes in the evaluation of risk means that its outcome may be different in kind or degree from QRA (see Chapter 7). The evaluation of risk is dependent on who the 'assessor' is (Hale, 1986). Furthermore, the overall process of risk assessment may be fundamentally different when applied to individual and social behaviour than when applied to the behaviour of engineering systems (Corrello, 1983). Hale (1986) has reviewed research into subjective evaluation of risk and has highlighted additional factors including:

1. The nature of the hazard;
2. The extent to which exposure to the hazard and its potential for harm are controllable;
3. The time scale over which any resultant harm may occur;
4. The assessor's knowledge and understanding;
5. The magnitude of the imagined consequences.

Human behaviour is generally not solely determined by the 'objective' estimation of risk as calculated by the numerical methods discussed in earlier chapters. In some cases the 'objective' estimation of risk will match our own 'subjective' evaluation and indeed may have played some part in determining our perceptions. Interestingly, the person can 'come into' the process of risk assessment at any of its many stages, and thus be given anything from a vague feeling of danger (pre-hazard identification) to a numerical risk estimate (post-estimation) to work with. Furthermore, human behaviour towards danger cannot be explained or predicted using a single measure of harm (see later). Beyond this, the collective perception of our reference group (the social perception) may also strongly determine our individual assessments, as, at an even higher level, may the public perception.

Often, public policy is determined more by collective perceptions of risk than by its more objective estimation (see Chapter 15). For example, the resources devoted to industrial safety far exceed those dedicated to road or home safety, yet compared to the latter the workplace is a relatively safe environment. There are about 450 deaths per annum in workplaces in the United Kingdom, while some 4500 are killed on the roads each year, and 6000 are killed through accidents in the home in the same period. Public reaction to safety is also more related to people's perceptions than to mere 'objective' fact. For example, public reaction to road deaths, occurring in many different accidents each involving a relatively small number of people, is very different to their reaction to air crashes. Although the latter involve more people per event, the total number killed per year is far fewer than on the roads. It would require, say, 20 jumbo jet crashes each year in the United Kingdom to match the death toll on our roads. Obviously to understand what is happening in anecdotal examples such as these, the perception of risk has to be systematically studied. Questions need to be asked.

The very nature of the question of the subjective evaluation of risk determines the type of methodology that must be used to derive not only the answers but also the theoretical context in which those answers are set. The methods that should be used to determine and structure perceptions of risk should be based on the individual and their self report of what they consider to be a risk and to what extent. This approach has been termed the 'Expressed Preference' approach (Royal Society, 1983). Such enquiries seek to understand not only the nature of hazards which are judged to be high-risk but the underlying decision making processes that accompany such judgements. Techniques available for this form of enquiry are derived from psychology, and the issue of risk perception is central to both applied and cognitive psychology.* It is from these domains that its theoretical basis is derived. Some researchers use a more behavioural approach to the area and consider the behaviour towards various hazards (Starr, 1968). This chapter will consider how we can proceed and further our understanding in this area.

Judgmental biases in risk perception

We start by considering the strategies people use when making judgements about hazards and their associated risks. For the lay person sufficient statistical evidence is seldom available and in the majority of cases inferences are made on the basis of incomplete information even when the person draws on their own existing knowledge (memory). Research has identified a number of general and simple inferential rules

---

* Cognitive psychology is the science of mental life and deals with processes such as perception, thinking and memory. Applied psychology is concerned with the application of psychological knowledge to practical problems.

that people use in such situations. These rules, known as 'heuristics', are employed here to reduce difficult mental tasks to simpler ones. The validity of different 'heuristics' is variable and can lead to large and persistent biases. We will later illustrate the nature of such biases by describing some of the relevant experimental studies. Tversky and Kahneman (1974) identified three heuristics commonly used in making judgements under conditions of uncertainty:

1.   *'Representativeness'* is used in judgements of the following type: What is the probability that object A belongs to class B? People who believe A belongs to class B may then assume that it shares the same risk. For example, is this chemical typical of other chemicals I have worked with, and thus poses the same risk?

2.   *'Availability'* refers to the ease with which an event or class of events can be brought to mind. People using this heuristic will tend to judge an event as likely or frequent if instances are easy to imagine or recall. However, availability of recall is also affected by numerous other factors (for example, exposure to media coverage). A vivid film or recent television programme could bias risk judgements.

3.   *'Adjustment and anchoring'* refers to the tendency to start from a preliminary value which is then adjusted to produce the final answer. A good example was provided by the early and developing debate over the hazards of working with visual display terminals. It was assumed early on in the VDT debate that the risk to all pregnant women was great, and since then the outcome of the relevant research has been serving to more tightly define the exact extent of the risk and the nature of 'at risk' groups.

For example, with reference to the 'availability' heuristic, overestimated causes of death, have tended to be dramatic and sensational, whereas underestimated causes have tended to be less spectacular events which actually claimed one victim at a time and were common in the non-fatal form (see later). This tendency could reflect the exposure to dramatic events through the media.

In a study of media coverage, Combs and Slovic (1979) examined the reporting of causes of death in two newspapers on opposite coasts of the United States over a period of time. As expected, violent and catastrophic causes of death were reported much more frequently than less dramatic causes of death with similar (or even greater) statistical frequencies. Such media coverage obviously provides a source of information for individuals and may be a contributory factor in availability bias.

The 'availability heuristic' also highlights the vital role of experience as a determinant of perceived risk. In particular, misleading experiences may underlie an individual's tendency to believe themselves to be personally immune to many hazards (*accidents only happen to others*). Road traffic accident research demonstrates how automobile drivers continue with unsafe driving behaviour because they make trip after trip without adverse incident.

In a study of the employee attitudes to safety in a major European Gas Company, one of the current authors had the opportunity to factor analyse attitude statements from 600 employees (Cox, 1988; Cox and Cox, 1991). A 'personal immunity' factor was identified among five groups of statements. The majority of those endorsing the constituent statements had not experienced an accident in their workplace.

Judgmental differences amongst individuals and groups

Both casual observation of current risk debates on nuclear energy and more systematic empirical data indicate that lay people and experts have different perceptions of risk associated with various technologies. Given the disparity of experience and knowledge, this divergence is not unexpected. Furthermore, research indicates that as evidence accumulates, public perceptions are slow to change and can be extremely persistent in the face of contrary evidence. Initial impressions about a hazard tend to structure the way that subsequent evidence is interpreted. It is therefore vitally important that information concerning hazards is communicated effectively from the start.

In a study of group perceptions, Lichtenstein (1975) and her co-workers asked different groups of lay persons (69 college students, 76 members of the League of Women Voters (professional women) and 47 business and professional members of the 'Active Club') and one group of experts (15 persons professionally involved in risk assessment in the United States) to judge 30 hazardous activities, substances and technologies according to the likely risk of death. Table 13.2 rank orders the mean risk judgement for the four groups. The lower rankings represent the most 'risky' activities, substances or technologies.

There were many similarities amongst the three groups of lay persons. In particular, each group judged the risk from motorcycles, motor vehicles and handguns as high, while vaccinations, home appliances, power mowers and football were seen to pose a lower risk. However, there were a number of interesting differences in the rankings. Nuclear power was rated highest by the League of Women Voters and the students, but only eighth by the Active Club. The students viewed contraceptives and food preservatives as riskier than the other two groups. However they tended to judge outdoor activities

Table 13.2 Ordering of perceived risk for 30 activities and technologies. (Rank 1 represents the most risky activity or technology)

	League of Women Voters	College Students	Active Club Members	Experts
nuclear power	1	1	8	20
motor vehicles	2	5	3	1
handguns	3	2	1	4
smoking	4	3	4	2
motorcycles	5	6	2	6
alcoholic beverages	6	7	5	3
general (private) aviation	7	15	11	12
police work	8	8	7	17
pesticides	9	4	15	8
surgery	10	11	9	5
fire fighting	11	10	6	18
large construction	12	14	13	13
hunting	13	18	10	23
spray cans	14	13	23	26
mountain climbing	15	22	12	29
bicycles	16	24	14	15
commercial aviation	17	16	18	16
electric power (non-nuclear)	18	19	19	9
swimming	19	30	17	10
contraceptives	20	9	22	11
skiing	21	25	16	30
x-rays	22	17	24	7
high school and college football	23	26	21	27
railroads	24	23	20	19
food preservatives	25	12	28	14
food colouring	26	20	30	21
power mowers	27	28	25	28
prescription antibiotics	28	21	26	24
home appliances	29	27	27	22
vaccinations	30	29	29	25

such as mountain climbing, skiing and hunting as safer than the other two lay groups. The experts' judgements differed from those of the lay groups on a number of items, most markedly in their ranking of nuclear power. In addition they viewed electric power, surgery, swimming and x-rays as more risky and police work and mountain climbing to be less risky than did the three lay groups.

The accuracy of judgements of 'experts' is not so clear in another study. The authors (Christensen-Szalanski et al., 1983) studied the estimates of mortality due to various causes made by a group of students and a group of doctors. Although the professional judgements tended to be relatively more accurate, they still showed substantial inaccuracies. There was a significant tendency for them to be influenced by recent experience, overestimating the risk of diseases they happened to have seen personally. Another interesting finding from this study was the fact that professionals showed the same patterns of influence as the public from the general social emphasis on a disease.

The age of those making the judgements is shown to be important. In a study on the influence of safety belt usage on perception of the risk of an accident in a group of young (under the age of 25) and older male drivers, the researchers found significant differences between the two groups (Tversky and Kahneman, 1974). Both groups were asked to drive on an urban route and to rate their perceptions of the risk of an accident. On the first driving trip all subjects were unbelted, while on the second trip half of the subjects wore a safety belt while half did not. Results showed that young male drivers decrease their perception of the risk of an accident as they become familiar with a driving route if they are not wearing a safety belt. Young male drivers asked to wear a safety belt sustained their perception of the risk of an accident as they became familiar with the test route. Older drivers' perception of the risk of an accident was not affected by familiarity or safety belt usage.

Factors which influence the perception of risk

In order to understand more about the underlying mental (cognitive) processes involved in the perception of risk, and to clarify what individuals mean when they classify specific activities or technologies (hazards) as 'risky', more sophisticated analytical techniques are necessary than those discussed in the previous section. Furthermore, factors other than mortality estimates are present when individual hazards are assessed.

In a further investigation of their earlier studies (see above) the researchers confirmed that lay persons' risk perceptions were based on more than fatality statistics and they initiated further studies to identify these considerations and to clarify the process. This study is described below (Slovic et al., 1984).

Availability

Availability bias has been demonstrated in a number of studies by Slovic et al., (1981). In their first study, subjects were asked to judge the number of deaths resulting from 40 different causes relative to the annual death toll in the United States (50 000) due to motor vehicle accidents. Figure 13.2 compares the judged number of deaths per year

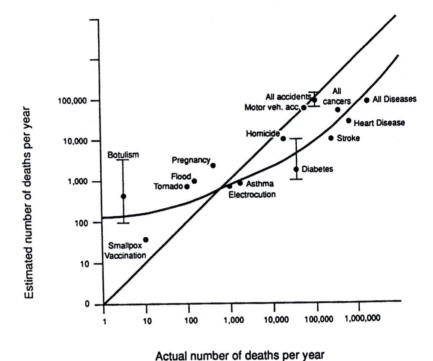

FIGURE 13.2
Comparison of number of judged deaths per year with actual deaths

with the number reported in official statistics. If their judgements equalled the statistical rates, all data points would fall on the identity line. However, the mean ratings were scattered about a curved line that lay sometimes above and sometimes below the line of accurate judgement. Several biases are evident in Figure 13.2. In general terms, rare causes of death were overestimated and common causes of death were underestimated. For example, accidents were judged to cause as many deaths as diseases, although in reality diseases cause 15 times the number of fatalities. Homicides were judged as more frequent than diabetes and stomach cancer.

It is also interesting to note the differences in spread of scores for a less common cause (botulism) compared to a more common cause (accidents) as indicated by the sample error bars. In a later study, 90 hazards representing a very broad range of activities were rated by a single group of lay persons (college students) according to 16 qualitative risk characteristics. The 16 risk characteristics are listed in Table 13.3 and were selected by the researchers to represent important areas of concern.

Each of the 90 hazards was rated on overall riskiness and judged

Table 13.3 Qualitative risk characteristics used in Slovic *et al.*'s factor analysis

1.	Uncontrollable	Controllable
2.	Dread	Non-dread
3.	Global catastrophic	Not global catastrophic
4.	Consequences fatal	Consequences non-fatal
5.	Not equitable	Equitable
6.	Catastrophic	Individual
7.	High risk to future generations	Low-risk to future generations
8.	Not easily reduced	Easily reduced
9.	Risk increasing	Risk decreasing
10.	Involuntary	Voluntary
11.	Affects me	Does not affect me
12.	Not observable	Observable
13.	Unknown to those exposed	Known to those exposed
14.	Effect delayed	Effect immediate
15.	New risk	Old risk
16.	Risks unknown to science	Risks known to science

on all 16 characteristics of risk. In general, the risks from most of these activities were judged to be increasing, not easily reduced, and better known to science than to those people exposed to them. Many of the qualitative risk characteristics were found to be highly correlated with each other across a wide range of hazards. For example, hazards rated as 'voluntary' tend also to be rated as controllable; hazards that threaten future generations tend also to be seen as having catastrophic potential.

Statistical examination of these interrelationships by means of factor analytical techniques showed that the 16 characteristics could be represented by two or three higher order characteristics or factors. Factor 1 was associated with lack of control, fatal consequences, high catastrophic potential, reactions of dread, inequitable distribution of risks and benefits and the belief that risks are increasing and not easily reducible. This factor was labelled 'Dread risk'. Factor 2 was associated with risks that are unknown, unobservable, new and delayed in their manifestation. It was labelled 'Unknown risk' (see Figure 13.3). Factor 3 was associated with the 'adverse consequences' and is not illustrated on this two-dimensional model (Figure 13.3).

The same researchers have subsequently reported studies which consistently replicate their psychometric results. Furthermore, they have found that lay persons' risk perceptions and attitudes are closely related to the position of a hazard within the factor space. Most important is the factor 'Dread risk'. The higher the hazard's score on this factor, the higher its perceived risk, the more people want to see its current risks

FIGURE 13.3
Plot of 'dread risk' against 'unknown risk'

reduced, and the more they want to see strict regulations employed to achieve the desired reduction in risk.

These studies have been the subject of lively debate in the scientific literature and other techniques have been used which allow respondents to generate their own responses, such as the repertory grid used by Green and Brown (1980) in work sponsored by the Fire Research bodies. Such techniques allow researchers to examine the way respondents interpret the nature of hazards in relation to a set of elements. Green and Brown have been particularly interested to see whether there are differences between 'immediate-in effect' hazards and 'delayed-in effect' hazards. They have done some work which included major chemical plants and nuclear plants. Interestingly correspondents appeared to see the nature of hazards from chemical plants as neither wholly delayed nor wholly immediate; certainly the responses to the perceived risks from the two types of plant were somewhat atypical, i.e. quite dissimilar to all other types of hazards, results which call into question the relevance of trying to compare chemical risks with risks such as travelling by car (Green, 1979).

The detailed study of single hazards

The use of comparative research strategies is pervasive in the literature, and they have even been used when the focus of interest is one particular hazard such as nuclear power. A major contribution here has been made by joint research groups sponsored by the International Atomic Energy Authority and the International Institute for Applied Systems Analysis at Laxenburg, near Vienna in Austria. While the primary aim has been to clarify public attitudes to nuclear power, their most definitive study does this by addressing five energy systems, nuclear, coal, oil, solar and hydro (Thomas, 1981). The research concentrates on the attitude measurement approach developed by Fischbein (1975). The rationale of the Fischbein approach to measurement can be explained fairly simply. A person's attitude to nuclear power, on the dimension from 'pro' to 'anti', in favour or against, can be desegregated into the set of beliefs they hold about the connections between nuclear power and a number of positive and negative attributes. These beliefs could be said to answer for people the question: what is nuclear power; what is it likely to do for us in terms of pleasant and unpleasant consequences? There is an implicit cost/benefit synthesis in the model. Three examples of belief statements which the subjects are invited to endorse or reject are 'nuclear power will increase employment'; 'nuclear power will produce changes in man's genetic make-up'; 'coal (mining) will exhaust our national resources'.

Factor-analysis of the beliefs leads to identification of four risk factors:

- psychological aspects;
- economic and technical benefits;
- socio-political implications;
- environmental and physical risks.

Factor 1: psychological aspects
- means exposing myself to risk without my consent,
- leads to accidents which affect large numbers of people at the same time,
- means exposing myself to risk which I cannot control,
- is a threat to mankind,
- is risky.

Factor 2: economic and technical benefits
- increases the standard of living,
- increases economic development,
- provides good economic value,
- increases my nation's prestige,
- leads to new forms of industrial development.

Factor 3: socio-political implications
- leads to rigorous physical security measures,
- produces noxious waste products,
- leads to diffusion of knowledge that facilitates the construction of weapons by additional countries,
- leads to dependency on small groups of highly specialized experts,
- leads to transporting dangerous substances.

Factor 4: environmental and physical risks
- does not exhaust our natural resources,
- increases occupational accidents,
- leads to water pollution,
- leads to air pollution,
- makes us economically dependent on other countries,
- leads to long term modification of the climate.

When the respondents are divided into 'pro' and 'anti' groups in terms of nuclear power, it is found that the 'antis' believe that the use of nuclear energy would lead to all four dimensions of risk, the strongest belief being in psychological aspects and environmental and physical risks. The pro groups do not believe that nuclear power is free from risk, although their concern is mainly restricted to that of psychological and physical risk, and here their conviction is less strong than that of the antis. The major difference lies in the perceived probable benefits of nuclear energy. The anti group just does not believe that there are significant technological developments to be gained and they have only a very weak belief in potential economic gains. The pros believe strongly in both.

Risk homeostasis and behaviour

It has been suggested that in some circumstances, people behave as if they aim to maintain a roughly constant level of risk. This theory has been termed 'risk homeostasis', and examples are usually given in relation to road accidents. The safer the construction of roads, the faster people drive, thus offsetting the reduction in risk brought about by improved design. The introduction of a requirement in Nigeria for motor cyclists to wear crash helmets resulted in those cyclists driving less safely in the belief that they were less vulnerable. Road deaths involving motor cyclists actually increased. If correct, the theory of 'risk homeostasis' may explain the failure of many safety interventions (Howarth, 1987).

However, whatever other criticisms are levelled at this theory, its underlying mechanism is based on 'perceived risk'. In both cases cited, behaviour became more risky because perceived risk was reduced. It may thus be essential in any safety intervention based on voluntary behaviour to maintain or accentuate perceived risk while reducing actual risk (for example, the introduction of seat belts while emphasizing the dangers of driving).

Implications of risk perception research

An understanding of how people think about risk has an important role in informing or educating people. It also has applications in the problem of understanding and forecasting public response to hazards. It can enable organizations based on technologies like nuclear power or genetic engineering to provide guidelines for managing the social conflicts surrounding hazardous technologies.

An important contribution of the existing research has been to demonstrate the inadequacy of the unidimensional indices (such as annual probability of death, loss of life expectancy) that have often been advocated for putting risks in perspective and aiding decision making. Psychometric studies have suggested that such comparisons are not totally satisfactory because individuals' perceptions are determined not only by mortality statistics but also by a variety of quantitative and qualitative characteristics. Other studies suggest that most individuals rely on partial information, imperfect memories and distorted time perspectives to extrapolate from past experiences into the future and that their consequent probability predictions are often inaccurate. Furthermore, there are gross differences in perception both between groups and individuals. Such differences are clearly demonstrated when one considers the risk perceptions of 'experts' in comparison to those of 'lay persons'. All of these factors highlight the need for continuing education and re-education.

Studies in the USA on various technologies (Lowrence, 1990) suggest individual perceptions of industrial risks mirror the following attitudes and beliefs:

1.   Fundamentally western industrialized countries are conceived as risk-buffering societies, reflected in 'compensation' schemes;
2.   Individuals strongly prefer to choose their own risks, they resent involuntary and imposed risk and extend this voluntariness to others;
3.   Individuals are willing to allow others to undertake risky actions if the consequences are internalized;
4.   Individuals are willing to condone a risk imposing activity if people are compensated;

5. Unspecified and undetermined consequences are not difficult to accept emotionally;
6. Catastrophic consequences are emotionally difficult to endure;
7. Individuals rate immediate consequences more highly than long-term ones;
8. Individuals have a need to delegate responsibility to both competent and trustworthy persons for assessing risks. If trust is lost it cannot easily be regained.

Clearly, risk questions are important areas of concern in a technological society and an overall understanding must accept that the individual's perception of risk and the risk assessments are different but complementary forms of rationality and we should work towards their synchronization. When new risks arise, they might be characterized in terms of the two-factor model (see Figure 13.3), and other existing risks occupying similar positions identified. Successful methods of dealing with the underlying hazards, based on manipulating the perception of risk, might then be transferred from the existing risk to the new one. This argument is further developed in Chapter 15.

Personal attitudes should be taken into account in communicating risk and in the further understanding and control of human behaviour. Various acceptance criteria form the basis of such communications (for example, in managing and siting of high-hazard operations).

Acceptance standards for QRA

The practice of QRA requires acceptance standards to be set and risk criteria are one form of these. Criteria for individual risk were developed in civil aviation from the late 1930s onward (see, for example, Tait, 1993). Farmer (1967) introduced the FN curve to display societal risk and proposed risk criteria for use in siting decisions for nuclear plant. Quantitative techniques are now being used in a wide range of situations as described in Chapter 12, and acceptance standards have come under increasing scrutiny.

Studies normally distinguish between tolerable and acceptable risks (HSE, 1992). There is a certain degree of risk which we are willing to tolerate even though it affords us concern, if we can see a benefit accruing. Under these circumstances we would wish to see risk reduced to a level that is as low as is reasonably practicable (ALARP). By contrast, a risk is acceptable if it is perceived to be at a sufficiently low level that we do not look for further reduction, although we may require assurance that the risk will not be allowed to increase from its present level. Figure 13.4 illustrates these points.

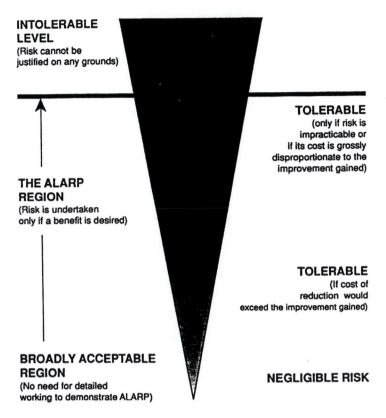

INTOLERABLE
LEVEL
(Risk cannot be
justified on any grounds)

TOLERABLE
(only if risk is
impracticable or
if its cost is grossly
disproportionate to the
improvement gained)

THE ALARP
REGION
(Risk is undertaken
only if a benefit is desired)

TOLERABLE
(If cost of
reduction would
exceed the improvement gained)

BROADLY ACCEPTABLE
REGION
(No need for detailed
working to demonstrate ALARP)

NEGLIGIBLE RISK

FIGURE 13.4
Levels of acceptable risk (HSE, 1988)

It is also important to distinguish between risk to the worker in the workplace and to members of the general public. The worker normally has a degree of choice about risk exposure and can see greater benefit to be obtained. The worker can thus be expected to have a tolerance to a higher level of risk than will members of the public.

The forgoing concepts can apply in the consideration of both individual and societal risk and the risk may at different times involve various categories of harm. In examining major accident hazards, death (either immediate or delayed) is frequently the category chosen. This was so in two HSE publications.

Individual risk

Tolerable and acceptable levels of risk are discussed in an HSE report first issued in 1988 and re-issued in 1992 (HSE, 1992). The levels of individual risk relevant to nuclear power stations and to other larger industrial installations were determined using available results of QRA

Table 13.4 Tolerable and acceptable levels of risk of immediate or delayed death

Nature of risk	Annual risk level	Other situations with similar risk
Maximum tolerable risk to workers in any industry	1 in 10^3	High-risk industries such as extraction of mineral oil and gas
Maximum tolerablerisk to members of the public from major industrial hazards	1 in 10^4	Risk of death in traffic accidents
Maximum acceptable risk	1 in 10^6	Risk of death by electrocution in the home

as a guide with additional cross referencing to risk levels in a broad range of other occupational and non-occupational situations. Employees and members of the public were considered separately. The levels suggested in the HSE (1992) report are reproduced in Table 13.4.

Societal risk

The same publication (HSE, 1992) considers societal risk. By examining the results of several assessments (see *FN* curves in Figure 11.3 for examples) they concluded that, for incidents resulting in hundreds or a few thousand deaths, the maximum tolerable societal risk will be of the order of 1 in 1000, or perhaps 1 in 5000 per year. The report points out that, in considering nuclear power, it is necessary to take into consideration all power plants contributing to societal risk. In the United Kingdom this will include contributions from power plants in neighbouring European countries. The report suggests that future nuclear power plants should be designed with a view to restricting overall probabilities to one incident involving 100–1000 immediate or delayed deaths per 10 000 years.

Another HSE publication, originally issued in 1989 and re-issued in 1993 (HSE, 1993), makes a detailed study of the 16 risk assessments covering a wide range of risk situations. The publication lists 41 factors which are considered important in judging the tolerability of societal risk. In the circumstances it is concluded that it is not possible to produce an upper envelope *FN* curve defining a societal risk which will be just tolerable in all situations.

Further reading

The Royal Society (1983) *Risk Assessment: A Study Group Report*, Royal Society, London.

The Royal Society (1992) *Risk: Analysis, Perception and Management*, Royal Society, London.
HSE (1988) *The Tolerability of Risk from Nuclear Power Stations*, HMSO, London.

References

Christensen-Szalanski, J. J. J., Beck, D. E., Christensen-Szalanski, C. M. and Koepsell, T. D. (1983) Effects of expertise and experience on risk judgements, *J. Appl. Psychology*, **68**, 278.

Combs, B. and Slovic, P. (1979) Causes of death: biased newspaper coverage and biased judgements, *Journalism Quarterly*, **56**, 837.

Corrello, V. T. (1983) The perception of technological risk: a literature review, *Technological Forecasting and Social Change*, **23**, 285.

Cox, S. and Cox, T. (1991) The Structure of Employee Attitudes to Safety: A European Example, *Work and Stress*, **5**, 93.

Cox, S. (1988) *Employee Attitudes to Safety*, MPhil thesis, University of Nottingham, Nottingham.

Farmer, F. R. (1967) Siting criteria – a new approach, *Nucl. Safety*, **8**, 539.

Fischbein, M. and Ajzen, I. (1975), *Belief, Attitude, Intention and Behaviour: An Introduction to Theory and Research*, Addison-Wesley.

Green, C. H. (1979) Risk: beliefs and attitudes in regard to major hazards, in *The Implications of Large Petrochemical Development*, Occasional Paper No. 1, Department of Town and Regional Planning, University of Dundee.

Green, C. H. and Brown, R. A. (1980) *Risk: beliefs and attitudes, in Fires and Human Behaviour*, D. V. Canter (ed.), John Wiley, Chichester.

Hale, A. R. (1986) *Subject Risk from Risks: Concepts and Measures*, W. T. Singleton and J. J. Harden (eds), John Wiley, Chichester.

HSE (1992) *The Tolerability of Risk from Nuclear Power Stations*, HMSO, London.

HSE (1993) *Quantified Risk Assessment: Its Input to Decision Making*, HMSO, London.

Howarth, C. I. (1987) Perceived risk and behavioural feedback: strategies for reducing accidents and increasing efficiency, *Work and Stress*, **1**, 61.

Lichtenstein, S., Slovic, P. and Fischhoff, B. (1975) Judged frequency of lethal events, *J. Exp. Psychol. (Human Learning and Memory)*, **4**, 551.

Lowrence, W. W. (1990), Stewardship of Chemical Production Risks. Paper presented at the First IUPAC Workshop on Safety in Chemical Production, September 9–13, Basle, Switzerland.

Royal Society (1983) *Risk Assessment: A Study Group Report*, London.

Slovic, P., Fischhoff, B. and Lichtenstein, S. (1981) Facts and fears: understanding perceived risk, in *Society Risk Assessment, How Safe is Safe Enough?* Richard C. Schwing and Walker A. Albers (eds), Plenum Press, New York, p. 181.

Slovic, B., Lichtenstein, S. and Fischhoff, B. (1984) Behavioural decision theory perspectives on risk and safety, *Acta Psychologica*, **56**, 183.

Starr, C. (1969) Social benefit versus technological risk, *Science*, **165**, 1232.

Tait, N. R. S. (1993) The use of probability in engineering design – an historical survey, *Reliability Engineering and Systems Safety*, **40**, 119.

Thomas, K. (1981) Comparative risk perception: how the public perceives the risks and benefits of different energy systems, *Proceedings of the Royal Society (London)*, **A376**, 35.

Tversky, A. and Kahneman, D. (1974) Judgement under uncertainty: heuristics and biases, *Science*, **185**, 1124.

Chapter 14

Risk assessment in occupational health and safety

Background and introduction

The use of risk assessment as an aid to the management of health and safety within organizations has increased significantly in the late 1980s and early 1990s as a direct consequence of new legislation. Although risk assessment was implicitly required in the United Kingdom Health and Safety at Work etc. Act, 1974, a number of subsequent pieces of legislation make it an explicit requirement. First, within the European Union the 1989 Directive 89/391/EEC made risk assessment a mandatory requirement. These requirements were then translated within the UK into the Management of Health and Safety at Work Regulations, 1992 (MHSWR). MHSWR introduced a general requirement for the employer to (i) undertake 'suitable and sufficient' risk assessments and (ii) introduce the necessary measures to control the hazards and limit the associated risks to 'acceptable' levels. The procedures used in compliance with these general requirements are described in outline in this chapter. They utilize much of the methodology and principles described in this text.

The 'language' of risk

Earlier chapters illustrated how instrumental failure to danger can lead to the introduction of hazard(s) in a variety of technical systems. For example, methods were described as to how:

- Such failure modes can be identified;
- Failure probabilities may be calculated;

- Outcomes and consequences predicted;
- The resulting risks quantified.

This process was referred to as risk estimation. The evaluation of the results of this process involves psychological processes such as risk perception and the judgement of tolerability or acceptability of risk (see Chapter 13). Together, the processes of risk estimation and risk evaluation form the components of risk assessment (see Figure 13.1) and may be applied to fulfil a 'suitable and sufficient' assessment of workplace risks.

When risk assessment is applied to the total workplace (rather than to a discrete piece of equipment) additional considerations must be taken into account. First, the estimation of risk is broader based. Although instrumental failure to danger may still be one important hazard source, many more hazards may arise from the varied and complex relationships between the worker, the technology and the working environment. Such factors as worker motivation and worker morale, standards of training and supervision and workplace ergonomics (see Chapter 7) may also have a very important (and sometimes detrimental) effect. Thus, new ways must be used to identify hazards, including in-depth analysis of work activities, examination of how jobs are organized and supervised and a study of individual tasks with a view to identifying the safety critical elements.

Estimation of outcomes and consequences is also more complex and in many cases a range of degrees of harm with differing likelihoods (or frequencies) can occur. Observed accident and incident rates may thus be used to support consequence predictions. Simple semi-quantitative scales of harm have been developed in some method-ologies (see, for example, Cox, 1992) and may be used by risk assessors. Such scales can provide valuable support to managers in the safety management process.

We have now discussed a number of features of the risk estimation process described in earlier chapters and have re-visited the associated terminology. However, it is also important to consider the overlap with the quantified risk assessment methodology used by the American Institute of Chemical Engineers (AIChemE, 1989) (see Chapter 12). The AIChemE methodology utilizes the term 'risk analysis' to convey the broader basis of risk estimation. This term is used in subsequent sections so as to provide continuity and to emphasize the overlaps in 'qualitative' and quantitative approaches to risk assessment.

Risk evaluation also needs to be modified to fit the requirements of MHSWR. The modified procedure will be developed to support decisions on whether present control measures are adequate or whether further measures are needed. Legislative requirements, which to some extent reflect public attitudes to the acceptability and tolerability of risk,

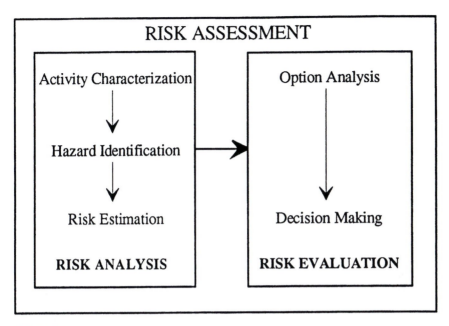

FIGURE 14.1
Risk assessment with its two components, risk analysis and risk evaluation

provide minimum standards. Health and safety standards and procedures within the organization and economic considerations centred on reasonable practicability will also be used as a necessary part of the evaluation process.

General methodology

Risk assessment, with its two components, risk analysis and risk evaluation, is illustrated in Figure 14.1. It is further described in the following sections. The description follows the processes and approach taken in the *Risk Assessment Toolkit* (Cox, 1992). These are outlined in Figure 14.2.

Analysis of work activities

The first stage in the process of analysing work activities is the 'walk through' survey. During this survey a note is made of the type of work undertaken in each area, the plant and equipment involved, an inventory of substances hazardous to health, responsible persons, and other relevant details. The walk through survey is essentially an analysis of work activities by geographical area. However, work activities may also be analysed in terms of generic activities (for example, use of display screens or work on mains electrical supplies, etc.). They may also be analysed by specific work tasks (for

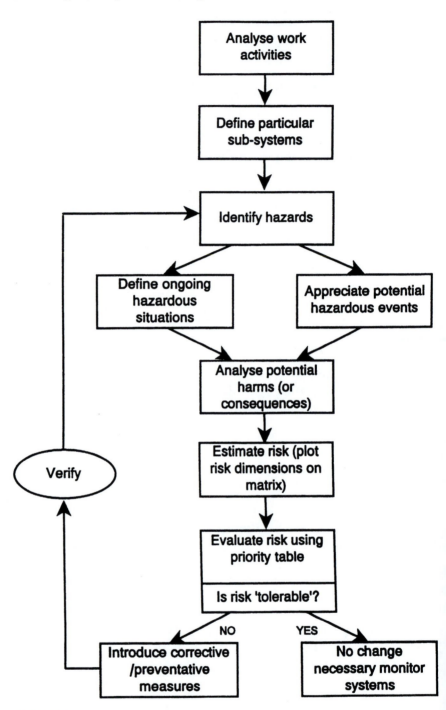

FIGURE 14.2
The risk assessment process (after Cox, 1992)

Table 14.1 Analysis of work activities

1.	Organizational analysis	
2.	Walk through survey	
3.	Analysis by	– geographical area
		– generic activity
		– specific work activity or substance

example, work in confined spaces or work involving lead). An organizational analysis may be carried out to support the 'walk through' survey. In this analysis, the assessor focuses on responsibilities in the workplace, standards of supervision, interrelationships (for example, shared areas and processes or the use of contractors).

Table 14.1 summarizes methods of analysing work activities. More details of the methods are to be found in *Risk Assessment Toolkit* (Cox, 1992).

Hazard identification

The second step in the risk assessment involves the identification of hazards. The Institution of Chemical Engineers (IChemE, 1985) has defined a hazard as a 'physical situation with a potential for human injury, damage to property, damage to the environment or some combination of these'. The Royal Society Study Group (Royal Society, 1992), in discussing hazards to people, provides the definition 'the situation that in particular circumstances could lead to harm'. The 'particular circumstance' aspect has been taken a stage further by one of the current authors (Cox and Cox, 1996), who introduces two further terms; the 'hazardous situation' in which a person interacts with the hazard but is not necessarily exposed to it, and the 'hazardous event' which triggers actual exposure of the person to the hazard.

Two examples providing a good illustration of the use of these terms are to be found in Table 14.2 (see Cox and Cox, 1996). The first involves nursing human immunodeficiency virus (HIV)-infected patients, the second, ascending and descending a staircase. The Table also suggests the most likely degree of harm to be expected in each case. The harm associated with falling down stairs will vary considerably from case to case and would be most serious where elderly people were involved. Categories of harm are discussed in more detail in Chapter 11.

A broad range of hazards and hazardous situations can be present in the workplace. These may be technical in origin, physical, chemical, biological, electrical or mechanical, for example, or may have ergonomic or psychosocial causes. A structured and systemic approach to hazard identification is essential if important hazards are not to be missed. Three general approaches are used (see Table 14.3):

Table 14.2 Hazards and harm – two examples

Concept	Example 1	Example 2
Hazard	HIV	Stairs
Hazardous situation	Nursing HIV patients	Ascending or descending stairs
Hazardous event	Needle stick injury causing contact with infected blood	Slipping or tripping on stairs leading to fall downstairs
Most likely harm	Delayed death within 15 years	Broken limb

- intuitive
- inductive
- deductive (Cox, 1992).

Brainstorming makes use of the intuitive approach. Participants in brainstorming should be selected from within an organization and should have as wide a range of relevant experience as possible. During the brainstorming process a free flow of ideas should be encouragedby setting a relaxed, non-critical atmosphere. Ideas are listed, consolidated, then further developed by the team. If used skilfully this technique can prove most effective.

Inductive methods focus on what could go wrong, or what might be expected to happen, in particular circumstances, given previous experience. Checklists and accident and occupational ill health statistics give valuable general guidance. Job Safety Analysis, in which a particular job is broken down into sub-tasks and each of these is investigated in order to predict where hazards might arise, is also a useful technique (see Cox and Cox, 1996).

The hazard and operability study (HAZOP) commonly used in the chemical industry has recently found a use in a broader context in planning safe procedures, for example in maintenance work. HAZOP is described in Chapter 4, as are failure modes and effects analysis (FMEA) and event tree analysis which are used to predict failure modes to danger in instrumentation systems.

Deductive methods start from what has gone wrong and use knowledge and experience to work back to 'deduce' the cause. In doing this, accident and incident databases can be very helpful. Fault tree analysis (see Chapter 4), starting from a 'top event', the hazardous outcome, predicts how such an outcome can be caused. Again, this technique is of particular relevance in instrumentation systems. Cox (1992) gives details as to how these hazard identification methods are used in practice. They are summarized in Table 14.3.

Table 14.3 Hazard identification methods

Method	Examples
Intuitive	Brainstorming
Inductive	Checklists
	Accident and occupational ill-health statistics
	Job safety analysis
	Hazard and operability study
	Failure modes and effects analysis
	Event tree analysis
Deductive	Accident and incident databases
	Fault tree analysis

Estimation of risk

The third step in the risk assessment process is the estimation of risk. The Royal Society report (Royal Society, 1992) defines risk as 'a combination of the probability or frequency of occurrence of a defined hazard and the magnitude of the consequences of the occurrence'. The Institution of Chemical Engineers (IChemE, 1985) gives the definition as 'the likelihood of a specified undesired event occurring within a specified period or in specified circumstances. It may be either a frequency (the number of specified events occurring in unit time) or a probability (the probability of a specific event following a prior event)'.

The 'per event' definition is used where causation is intermittent. For example, one might specify the risk per landing that an aircraft instrument landing system (ILS) will fail causing loss of life. The duration of the flight is irrelevant – the hazard arises only when the landing is attempted. In many cases the hazard is of a continuous nature and the 'per unit time' definition is used. If we consider the risk of a person working beneath scaffolding on a construction site receiving a fractured skull from an object falling from above, the risk per object falling is of no particular significance. The risk per day or per year is now a more appropriate measure.

Estimation of overall risk can present problems. In our first example, the ILS will have been designed to attain a certain (very low) failure rate, while accident and incident reports will indicate the proportion of ILS failures resulting in fatality. The probabilities are combined by multiplication:

The risk per landing of ILS failure leading to fatality $=$ Probability per landing of ILS failure $\times$ Probability of the resulting loss of control leading to fatality

In fact, ILS is designed under airworthiness requirements for this risk to be not more than 10^{-7} per landing. In the second example, published accident statistics may well provide a risk value directly. In the UK the HSE publish such statistics annually and it is possible to look up the number of fractures to the skull of employees on construction sites. This would provide an upper limit – it would be necessary to find out how many of the fractures were associated with objects falling from scaffolding. An alternative approach is to estimate the component probabilities. Thus:

| The risk per worker per day of skull fracture | = | Probability per day of object falling from scaffolding | × | Probability that worker is struck by it | × | Probability that, if struck, skull will be broken |

Note the extra factor this time, as we are dealing with risk per unit time. The third factor can almost certainly be determined from accident statistics and reports. If we can allocate even quite approximate numbers to the other two, a rough risk estimate can be made. We return to this approach shortly.

The definitions of risk quoted earlier involve two independent factors or dimensions. One is the probability or likelihood, the other is the severity of harm or consequence. Since these factors are independent, we can display them on a risk matrix. This is seen in a simple form in Figure 14.3, based on Cox (1992). Here we have divided both likelihood and consequence into three categories – 'low', 'medium' and 'high'. Thus, the upper of the two marked elements in Figure 14.3 is medium consequence/high likelihood. This approach can be taken a little further by allocating 'scores' of 1 to 3 and providing approximate ranges of consequence and likelihood as in Table 14.4. We can also increase the number of categories to four or five or more.

The risk analysis process described here is itself very useful in that it introduces a systematic approach to the hazards encountered in the workplace and their associated risks. Risk evaluation allows us to take a further step – we can examine the adequacy of the measures we have in place to control the hazards, and in cases

Table 14.4 A 3-by-3 risk matrix for occupational injury

	Low	*Medium*	*High*
Score	1	2	3
Consequence	Death or major injury	More than 3 days off work	First aid required
Likelihood	Not more than monthly	Every week or so	More than once per week

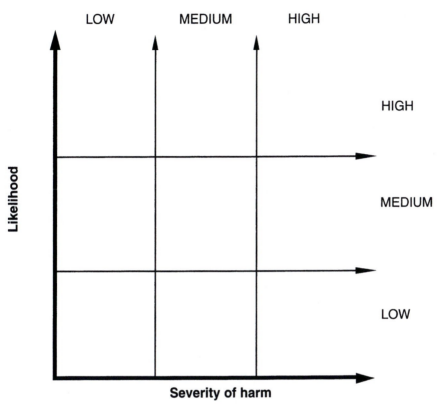

LOW   MEDIUM   HIGH

HIGH

MEDIUM

LOW

Likelihood

Severity of harm

FIGURE 14.3
The risk matrix (Cox, 1992)

where the risks are considered excessive, to determine priorities
for improvement.

Risk evaluation

The risk evaluation process would be greatly simplified if we
could develop a measure, expressed in terms of the two factors used
to define mathematical risk, which would provide a workable
representation of perceived risk. There has been some discussion about
this (Kaplan and Garick, 1981; Cox *et al.*, 1993) but practical experience
indicates that, in many situations, the product of the two factors
provides an adequate basis for prioritization, at least where common
hazards are involved. Applying this to our three-by-three matrix in
Figure 14.3, two elements would each have a score of 6 (3×2 and 2×3),
the maximum score would be 9 (3×3) and the minimum 1 (1×1).
However, there is still a significantly different risk profile for a low
consequence/high likelihood and a high consequence/low likelihood
situation which argues for a conservative approach to evaluation of

simple mathematical models.

Steel (1990) has described two more complicated schemes which employ the same multiplicative approach. In the first of these the matrix is extended and the likelihood dimension is evaluated from two factors which represent (likelihood of happening) and (frequency of task). This is explained in Table 14.5 where it is seen that consequence is derived from a simple table, likelihood being on a scale 1–8, consequence on a scale 2–10. High, medium and low risk are defined by different areas of the risk matrix with only slight numerical overlap and the associated actions levels are defined as follows:

High risk   Action within seven days. If not practicable, proof of steps taken to implement must be shown and written procedures must be used immediately. If not practicable to reduce risk, activity only* to be undertaken by highly trained specialist personnel.

Medium risk   Action plan to reduce risk to be drawn up. Until risk reduced, written procedure required. Reduce if reasonably practicable. Only trained personnel

Table 14.5 Matrix-based risk assessment (after Steel, 1990)

Consequence	
Death	10
Major injury	9
Long term sickness	8
Greater than 3 days	6
Less than 3 days	4
Minor injury	2

Likelihood of happening	Frequency of task		
	Frequent	Occasional	Hardly ever
Common occurrence	8	6	4
Frequent occurrence	6	4	3
Occasional occurrence	4	3	2
Improbable occurrence	2	1	1

Action levels

	10	9	8	7	6	4	2	
8	80	72	64	56	48	32	16	
6	60	54	48	42	36	24	12	
Likelihood 4	40	36	32	28	24	16	8	MEDIUM
3	30	27	24	21	18	12	6	
2	20	18	16	14	12	8	2	
1	10	9	8	7	6	4	2	LOW

HIGH (above)   CONSEQUENCE

Low risk
to undertake task. Written procedure required.
Reduce if practicable. Ensure personnel are competent to task.

In Steel's second scheme (Table 14.6) three factors are multiplied together to provide a measure of likelihood:

$$\text{Likelihood} = \frac{\text{Probability of exposure}}{\text{to or contact with hazard}} \times \frac{\text{Frequency of}}{\text{exposure to hazard}} \times \frac{\text{Number of}}{\text{persons at risk}}$$

Consequence (termed maximum probable loss) is defined in a similar way as previously and the perceived risk is represented by a hazard rating number (HRN) such that:

$$[\text{HRN}] = [\text{likelihood}] \times [\text{maximum probable loss}]$$

Action levels are directly related to HRN (Table 14.6).

Raafat (1995) has introduced a graphical system originally intended for the management of machine safety but now more widely used. Risk level is determined graphically from probability level, percentage of time exposed to the hazard and consequence level. Both Cox (1992) and Raafat (1995) include scales of consequence for economic harm and harm to the environment as well as harm to people. Raafat's scales are given in Table 14.7.

The semi-quantitative methods used by these various authors promote a systematic, structured approach to the assessment of risks in the workplace and provide a valuable aid to decision-making. All are in extensive use and typical examples of their application are to be found in Raafat (1995), Walker and Dempster (1994) and Walker and Cox (1995).

Published guidance

There are a number of sources of guidance on practical risk assessment within the United Kingdom. These range from the 'Five steps to risk assessment' leaflet (HSE, 1994) to the section on risk assessment published in the British Standard 8800 (BSI, 1996). The approaches described in these publications are similar to those of the current authors. Both publications provide practical guidance on implementing risk assessment which can be used to support the implementation of MHSWR.

The European Commission has also produced guidance on risk assessment at work (European Commission, 1996). The purpose of this guidance is also to help Member States and management and labour to fulfil the risk assessment duties laid down in framework directive 89/391/EEC. The Commission has also included a section

Table 14.6 Risk assessment based on hazard rating number (after Steel, 1990)

Probability of exposure to/contact with hazard		Frequency of exposure to hazard		Maximum probable loss (work related)		Persons at risk	
Exposure/contact	PE value	Frequency	FE value	Loss	MPL value	Number	NP value
Impossible	0	Infrequently	0.1	Fatality	15	(1–2)	1
Unlikely	1	Annually	0.2	Perm major illness/injury	8	(3–7)	2
Possible	2	Monthly	1.0	Temp major illness/injury	4	(8–15)	4
Even chance	5	Weekly	1.5	Major illnessi/injury	2	(16–50)	8
Probably	8	Daily	2.5	Minor illness/injury	1	(50>)	12
Likely	10	Hourly	4.0	Puncture wound	0.5		
Certain	15	Constantly	5.0	Scratch/bruise		0.1	

HRN	Risk	Agree action requirement to reduce HRN
(0–1)	Acceptable	Accept risk
(1–5)	Very low	Within 1 year
(5–10)	Low	Within 3 months
(10–50)	Significant	Within 1 month
(50–100)	High	Within 1 week
(100–500)	Very high	Within 1 day
(500–1000)	Extreme	Immediately
(>1000)	Unacceptable	Stop the activity until risk substantially lower

Table 14.7 Consequences of exposure to hazard (after Raafat, 1995)

Consequence level	1	2	3	4	5	6
Personnel	Insignificant	Minor	Major	Severe	Fatality	Multi-fatality
Economic	<£1000	<£10 000	<£100 000	<£1million	>£1million	Total loss
Environment	Minor	Short-term	Major	Severe	Widespread	Catastrophe

specifically aimed at the needs of small and medium sized enterprises (SMEs).

References

AIChemE (1989) *Guidelines for Chemical Process Quantitative Risk Analysis*, Centre for Chemical Process Safety of the American Institute of Chemical Engineers.

BSI (1996) *Guide to Occupational Health and Safety Management Systems*, British Standards Institute, London.

Cox, S. J. (1992) *Risk Assessment Toolkit*, Loughborough University.

Cox, S. and Cox, T. (1996) *Safety Systems and People*, Butterworth-Heinemann, Oxford.

Cox, T. and Cox, S. J. (1993) *Psychosocial and Organizational Hazards: Monitoring and Control*, European series in Occupational Health No. 5, World Health Organization.

Cox, T., Ferguson, E. and Farnsworth, W .F. (1993) *Nurses' Knowledge of HIV and AIDS and their Perceptions of the Associated Infection at Work*. Paper to the VI European Congress on Work and Organizational Psychology.

European Commission (1996) *Guidance on Risk Assessment at Work*, EC, Luxembourg.

HSE (1994) *Five Steps to Risk Assessment*, IND(G)163L, HMSO, London.

IChemE (1985) *Nomenclature for Hazard and Risk Assessment in the Process Industries*, Institution of Chemical Engineers.

Kaplan, S. and Garick, B. J. (1981) On the quantitative definition of risk, *Risk Analysis*, 1, 11.

Raafat, H. (1995) *Machine Safety – the Risk Based Approach*, Technical Communications (Publishing) Ltd.

Royal Society (1992) *Risk: Analysis, Perception and Management*, The Royal Society, London.

Steel, C. (1990) Risk estimation, *Safety and Health Practitioner*, June, 20–24.

Walker, D. and Cox, S. J. (1995) Risk Assessment: training the assessors, *The Training Officer*, July/August, 179–181.

Walker, D. and Dempster, S. (1994) *The implementation of a risk assessment programme: a case study*, Ergonomics and Health and Safety, The Ergonomics Society, Proceedings of Meeting, College Green, Bristol, September.

Chapter 15

Risk management and communication: decision-making and risk

Dr Judith Petts, CHaRM, Loughborough University, UK

Introduction

Earlier chapters have explored the development and use of quantified risk assessment (QRA) and probabilistic risk assessment (PRA) in the nuclear industry as an aid to the design and management of engineering systems. QRA has become an essential part of decision-making processes concerned with the location and control of technology in the public environment. The use of QRA in this context is a subject about which diverse views are held by industry and decision-making authorities alike. By generating numerical values for the consequences and probabilities of adverse events, QRA can bring objectivity to the decision-making process. However, QRA is not, and never will be, a precise scientific method and should not be seen as a mechanistic or automatic means of making risk management decisions. Whilst this should not undermine its usefulness for specific applications, the decision-making process has to be able to deal with risk in the context of public acceptability and concern over issues of equity, efficiency and consent. In this context, social and economic considerations will be as important as technical considerations.

QRA can be an invaluable tool in the communication of risk to decision makers and the public, most particularly if the assessment has included an understanding of public perceptions and concerns in the development of the criteria it has used to determine acceptability. However, successful risk communication programmes will be based upon an understanding of the decision-making process itself. The uncertainties in the risk assessment and the basis upon which conservative judgements have been made; an awareness of the 'hidden agendas' which have influenced the risk assessment and which may influence the risk decision; and an understanding of the costs and benefits of the different decision options. Risk communication is a fundamentally important part of the whole risk management process. Ineffective communication can lead to ineffective decisions. This chapter considers the effectiveness of risk communication in decisions where public and environmental safety issues are important. First, however, it is appropriate to explore further the concept of risk management and to consider the role and nature of risk communication within this process.

Risk management

Risk management is the term normally applied to the whole process of risk identification, estimation, evaluation, reduction and control. It can be considered to have at least six interlinked phases (see Figure 15.1), each incorporating a potential number of actions according to the 'project' to be managed:

1. *Hazard identification* – scoping the sources and components of a hazardous event, including targets which could be at risk.
2. *Hazard analysis* – determining release probabilities and rates, pathways of release and fate of substances in the environmental media in which they move and estimation of concentration at targets at risk.
3. *Risk estimation* – quantitative analysis of toxicological or epidemiological data; estimation of levels of human exposure; dose–response extrapolations; assessment of probabilities.
4. *Risk evaluation* – judgement of the significance of assessed risks; risk benefit analysis; risk acceptability; public perceptions of risks; economic impacts; uncertainty in risk estimation.
5. *Implementation* – development of implementation strategy; examination of policy options; siting decisions; plant design and layout; implementation of quality systems.
6. *Monitoring and auditing* – environmental monitoring; operations auditing; prospective epidemiology; new health risk information.

The 'project' could be the siting of a new hazardous installation or of development in the vicinity of a plant or siting of a new waste treatment or disposal facility, production of a medicinal compound, transport of dangerous chemicals, siting of an airport, etc. Phases 1–4 inclusive are often referred to as *risk sssessment*, phases 4–6 inclusive as *risk reduction*.

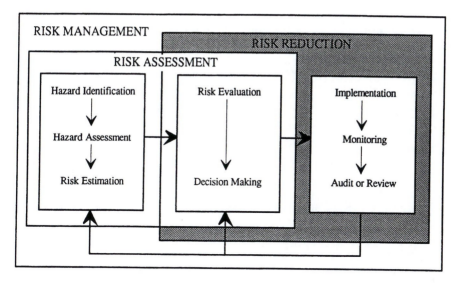

FIGURE 15.1
Risk management system

Formal models have been developed by several national and international agencies to describe the risk management process, including the Royal Society (Royal Society, 1983; 1992); the National Research Council in the USA (National Research Council, 1983); the World Health Organization (WHO, 1985); the Interdepartmental Working Group in Canada (Interdepartmental Working Group, 1984). These models exhibit many similarities, and serve to clarify the important elements of risk management.

The purpose of risk management can be identified in the following aims:

1. To control and reduce risks to acceptable levels;
2. To reduce uncertainty in risk decision-making;
3. To increase the public credibility of risk management decisions.

Given these aims, it can readily be seen that probably the single most important element of risk management is the transfer of risk information (or risk communication) between those measuring the risk and those who have to make decisions (formal and informal) about the risk.

Risk communication and risk management

Risk communication can occur at all of the stages of the risk assessment/risk management process (Vertinsky and Vertinsky, 1982). Indeed, communication of risk information is a critical component of the whole process. The Royal Society (1992) referred to the study and practice of risk communication as 'a relatively new development'. In fact, risk communication discussion has developed over a period of some 20 years. Initially focused on the technical assessment of risk, understanding has developed to consider communication as a process of involving people in decisions (Fischhoff, 1995).

'Risk communication' has been formally defined by Covello *et al.* (1986) as 'any purposeful exchange of information about health or environmental risks between interested parties'. This definition, supported by the US National Research Council (1989), stresses communication as a two-way rather than a one-way process of information provision from 'expert' to any other party.

The interested parties in the risk management process could include any, or all, of the following:

- Government/regulatory agencies
- Corporations and industry (both groups and individual companies)
- Trade unions
- The media
- Scientists and independent experts
- Professional organizations and institutions
- Public/environmental interest groups
- Local community/action groups
- Individual citizens.

Table 15.1 identifies some of the main decision-making activities and 'actors' where risk communication is often important. These activities primarily involve the implementation and evaluation of regulations. The 'actors' include those responsible for regulatory decision-making, those to be regulated, those whose interests must be protected, particularly consumers and the general public, and those responsible for transmitting and shaping risk information (O'Riordan, 1985). The decision-making activities involve four characteristics common to all forms of risk regulation and management in different political and cultural traditions:

1. *Consultation* – some form of dialogue between key affected parties
2. *Dependence on expertise* – whilst expertise is vital, changing attitudes to 'expertise' and new forms of participation and consultation have changed the nature of the expert's role
3. *Self-regulation* – self-policing by risk creators can supplement agency resources in monitoring and can lead to sharing of expertise
4. *Political considerations* – risk regulation and management, once

Table 15.1 Main decision-making activities and actors where risk communication is important

	Activity	Actor
1.	Setting safety and environmental standards	Regulatory agency – using expert advice often response to public pressure
2.	Strategic planning	Local authorities; regulatory agencies; local communities; industry; media
3.	Siting decisions	Local authorities; regulatory agencies; local communities; interest groups; individuals; industry
4.	Licensing authority plant or product	Regulatory agency; industry
5.	Monitoring – process, product, safety; environmental quality	Regulatory agency, industry, independent consultants; media; victims
6.	Enforcement	Regulatory agency; industry; courts
7.	Emergency planning	Emergency authorities; industry; local communities
8.	Evaluation of effectiveness of regulation	Government; regulatory agency; public; media

primarily a technical activity, must now involve consideration of political and social factors. Thus, we often see selection of the technical option displaying a level of 'acceptable' risk which is politically and socially palatable, although from the expert's viewpoint it may not technically be the most attractive option (O'Riordan, 1985).

Different stages of the risk management process will involve different forms, levels, and objectives of risk communication. For example, during the hazard identification stage, risk communication could be in the form of the circulation of findings following a case study of an accident, or a working group exchange of views or information during a fault tree analysis. Importantly, the flow of information during this phase of the risk management process is likely to be between experts, the purpose being to ensure comprehensiveness of input to the analytical process and exchange of expert opinions and views.

This will be very different to the risk communication process during the risk reduction phase. Here the flow of information is likely to be between experts and/or regulatory agencies, and the public and/or local decision-making authorities. The objective of the risk communication process may be to reassure the public as to the safety of the project or the soundness of the data used, to provide an opportunity for a local community to increase their understanding of the operations of a facility, to seek to change attitudes or behaviour, or to alert or arouse people as to the actions to be taken in the event of an emergency. Otway and Wynne (1989) have referred to the reassurance element of risk communication in relation to siting decisions and the arousal element in relation to emergency planning.

Risk communication models

A number of models have been developed to describe the risk communication process, each primarily borrowing from the traditional source-receiver model of communication (Lasswell, 1948):

1. *Information flow model* (Baram, 1984)
This model focuses on the movement of information from those who initially process it to those who ultimately require it. The model is rooted within a legal framework in which responsibilities and liabilities can be assigned to the various actors on the basis of the risk information which they transmit to others. The model treats the general public and interested parties as passive recipients of information overlooking the important elements of understanding and perception.

2. *The message transmission model* (Covello et al., 1986)
This model uses the engineering theory of communications, treating the message as an electronic signal and looking at the capability of the system to reproduce this signal without distortion at the receiving end. Thus, the model focuses on the problems of communication including:

(a)   Message problems – for example, deficiencies in knowledge and scientific understanding;
(b)   Source problems – for example, disagreements between experts; resource limitations which prevent reduction of uncertainty; use of technical or legalistic language leading to a lack of trust and credibility in experts;
(c)   Channel problems – for example, biased media reporting; premature disclosure of information; inaccuracies in interpretation of information;
(d)   Receiver problems – for example, lack of interest, misunderstanding of evidence, unrealistic expectations about the effectiveness of regulatory action.

The strengths of this model lie in its identification of the difficulties at every stage in the communication of information, and the fact that it identifies the process of communication as a dynamic process. However, a noticeable weakness of the model is that it still views risk communication as a collection of one-way transmissions of information.

3. *The communications processes model* (Gregory, 1989)
This model seeks to use the best features of the previous models and to improve them. The model recognizes two areas – 'technical' risk and 'perceived' risk. It seeks to explain the 'actors' in terms of the language each normally uses. Thus, industry and experts are in the area of technical risk, whilst the media and the public are in the area of perceived risk, and finally the government is in a position between the two and thus needing to communicate in both areas. The model identifies that most communication problems arise with respect to communication between the two areas, rather than with communication within an area. The model thus stresses the tension between technical and perceived risk.

4. *The social amplification model* (Kasperson *et al.*, 1988; Renn, 1991)
This model addresses the range of psychological, social or cultural processes which interact to either intensify or alternate perceptions of risk. A hazardous event is considered as an outgoing message which is 'filtered' through a series of 'amplification stations', scientists, the media, government agencies, activist groups, etc., each either amplifying or attenuating the risk message. The model servers to stress that risk must be viewed from a multiple perspective, with all messages about risks subject to 'noise'.

Perceived and acceptable risk

Chapter 13 examined judgements about the significance of hazardous events as perceived by individuals and by the various groups of individuals which make up society. It became clear that human behaviour is not solely determined by the 'objective' estimation of risk as calculated by the various numerical methods described in earlier chapters. Importantly, Chapter 13 concluded that personal attitudes should be taken into account in communicating risk and in the further understanding and control of human behaviour. Psychometric studies have aided the understanding of *why* people might be concerned about industrial or waste disposal activities, for example, but further analysis is required if we are to understand how to improve these perceptions.

Sandmann (1988) explained people's concerns and fears as a product of outrage rather than of hazard. People tend to overestimate hazard, and resist it vehemently, when the outrage is high. Conversely, they will underestimate hazard and respond apathetically when the outrage is low. Sandmann, building upon earlier discussions of conflicts over waste facility siting (for example, Hirschhorn, 1984), identified the public's lack of trust in responsible authorities and industries as one of the important sources of outrage.

The risk management literature (for example, Renn and Levine, 1991) has defined the components of trust to include:

- Perceived competence
- Objectivity (lack of bias)
- Procedural fairness in decision-making
- Consistency
- A perception of 'goodwill' in composing information.

Importantly, 'who' is communicating appears to have an important influence upon public trust, with industry and government representatives scoring particularly badly (Covello, 1992; Marris *et al.*, 1996).

Cvetkovich and Wiedemann (1988), considering how to optimize trust, identified the need to tailor risk communication strategies according to the specific problem. Engineering component choices in a safety system might be considered as a low level problem, one where the technical issues can be solved by the appropriate transmission of information. The issue the risk communicator is facing is the public's confidence in the given information. The risk analyst has to demonstrate that in gathering the information he has applied the proper rules. The criterion that determines the audience's confidence is evidence. This contrasts with higher-level problems such as the management of radioactive waste or worldwide problems such as the depletion of the ozone layer. Here the problems involve issues of value considerations and ethical and moral questions. The issue the risk communicator is facing is the public's trust. Decision-making has to be by means of debate and argument to create shared understanding and meaning.

Decision-making/decision analysis
The potential for the public to be 'sensible about risk' given 'sufficient time to reflect upon balanced information' remains the hope of many decision makers (HSE, 1996). The deficit model of the public as knowledge deficient and misguided supports this expectation (Wynne, 1991). However, the reality is more complex than this. Often the public may understand perfectly well what is being said, but they still disagree with the 'expert' because they are starting from a different value judgement position. Poor communication may be evident, but risk acceptance disputes often reflect more fundamental differences in the way risk allocators deal with scientific and societal uncertainty, and the way that the different communities at risk view these questions.

The deficit model supports a view of 'experts' and scientists as rational, making objective judgements about risk. Wynne has stressed that this over-simplistic view fails to acknowledge the fact that experts make social judgements and valuations which define how a risk system is to be analysed. These assumptions are often an unconsciously expressed function of their own social roles and relationships within the system (Wynne, 1989a). Importantly these value assumptions have not always been openly expressed and, thus, explored.

When units are attached to quantitative risk assessments many value choices are made, whether or not these are explicit. Choosing the right index of risk for any given decision process is a value-laden art (Watson *et al.*, 1984; Wynne, 1989b). Whilst this 'art' cannot be avoided it should be as obvious, considered, and subject to public involvement as is possible. Decision makers rather than experts should choose which attributes (as identified in risk perception work) to use, with what weighting. For example, if delayed death is evaluated more seriously than immediate death, or injury to children is considered to be more serious than a few years general life shortening, this should be openly identified. Having performed such an evaluative exercise in an explicit way for different options the decision maker should be able to trade off the benefits and risks to arrive at a choice which, given enough accountability and consensus, should be the society's choice.

Risks, costs and benefits are often distributed unevenly through a society (local, regional and national). In order to be acceptable, the risk management process must be able to consider issues of equity, or fairness, in the distribution of these risks and benefits. Four equity factors, or considerations, have been described as the 'principles of justice' (Kasperson and Kasperson, 1984):

1. *Utility.* The aim should be to maximize the welfare of all people in the society who are judged to be relevant to the activity to be controlled.
2. *Ability.* The ability of different members or groups of the society to bear risks should be considered.
3. *Compensation.* If necessary, compensation will need to be paid to those who have to bear an element of residual risk.
4. *Consent.* The informed consent of those who are affected or have to bear the residual risk is necessary.

Most of the risk communication literature had, by the early 1990s, begun to focus on the need to optimize consensus-building processes in decision-making (Renn, 1991; Renn *et al.*, 1995). Consensus-building seeks to improve the quality of public participation in decisions by (a) effective empowerment of the public, (b) a fair decision, and (c) active support of the final decision as being the best that can be achieved in the circumstances.

As a means of exploring the potential for improving equity in decision-making, there follows an examination of risk communication in the United Kingdom in relation to land-use planning, major hazard control and waste incinerator siting. Both decision-making areas have seen the development of use of QRA, albeit with differing pressures.

In Europe, the United Kingdom and the Netherlands have particularly developed the use of QRA in siting decisions relating to major hazards. It is noticeable that both countries have well-developed and formal land-use planning systems. This compares, for example, with the United States where planning control is a far more recent phenomenon with variable adoption at state level. Whilst QRA has dominated United States regulatory decision-making in relation to chronic risks to public health, particularly from carcinogenic substances, its use in siting decisions has been less developed.

In Britain, the use of QRA and the publication by the regulatory agency the Health and Safety Executive (HSE) of a document on acceptable risk criteria in land-use planning decisions for development in the vicinity of major industrial hazards (HSE, 1989a) has been part of a specific risk communication exercise. The use of QRA by the HSE in relation to land-use planning for major hazards and publication of acceptable risk criteria mirrors the use of Probabilistic Risk Assessment (PRA) in relation to the siting and control of nuclear power installations and the publication of 'tolerability' criteria (HSE, 1992).

The publication of the HSE's 'tolerability' document was in response to Sir Frank Layfield's Report on the Sizewell B public inquiry (Layfield, 1987). The latter concluded that 'the opinion of the public should underlie the evaluation of risk...there is at present insufficient public information to allow understanding of the basis for the regulation of nuclear safety'. The publication of the standards of safety implicit in the licensing and control of nuclear installations was a major step forward in the United Kingdom in terms of open discussion of a regulatory agency's control standards. Against a background of public concern, which had been witnessed at the Sizewell inquiry, the document sought to show what levels of risk in civil nuclear regulation might reasonably be regarded as tolerable in comparison with other risks in life. The document stressed that tolerable should not be equated with acceptable. Rather, it refers to a willingness to live with a risk so as to secure certain benefits and in the confidence that the risk is being properly controlled.

The HSE followed publication of the nuclear tolerability document with a public report describing the use of QRA (and PRA) in decision-making (HSE, 1989b). The report stresses that any decision process involving QRA (whether for control of exposure to radiation, exposure to certain dangerous substance, or siting of hazardous installations) must combine:

1. Quantification of likely risk with an understanding of the inherent uncertainties in this (essentially technical);
2. Reference to the benefits generated by the project and the political and economic considerations associated with it;
3. Weighing of what might be judged tolerable or intolerable by the public;

4.   A decision as to how far further reduction of risk could reasonably be attempted, taking cost into account.

The development and use of QRA in relation to waste incinerator siting has had a different history to that of major hazards. In the United Kingdom, practice preceded regulatory attention as public concerns about siting, in particular about the health risks of dioxins, placed pressures upon proponents of new facilities to undertake QRA of emissions. The first assessments were incorporated into Environmental (Impact) Assessments (Petts and Eduljee, 1994). Since the late 1980s, use of QRA in this respect has continued to develop in an *ad hoc* manner without any pressure, or indeed assistance, from the regulatory authorities (unlike for major hazards). Since 1989, at least 18 environmental impact statements accompanying planning applications for new incineration capacity have included a full quantified health risk assessment.

Political and regulatory encouragement of the use of risk assessment (Department of the Environment, 1995a) is now based on a view that it provides for a non-prescriptive regulatory approach which also meets public demands for evidence of control through a focus on local, site-specific risk decisions. Importantly, the use of QRA in environmental safety decisions such as the siting of incinerators focuses on chronic risks from ongoing emissions as well as acute risks arising from storage, plant failure, transport, etc. The following discussion reveals, however, that many of the risk communication issues are similar.

The discussion focuses on a number of important issues in the exploration of risk communication strategies:

1.   Risk management is (or should be) a process of bargaining in which QRA can play an important role in discussion of what risks are acceptable, particularly in the local context. However, the management process also has to involve some discussion as to how the costs of controlling the residual risks can most appropriately and equitably be met. Risk communication which aims only to persuade or promote understanding of the hazards and risks may not be effective in this process.
2.   Risk communication has to be a two-way process between regulatory agencies, local communities and industry. But do the decision-making frameworks provide for such a process?
3.   Questions of risk acceptability and control are location, social context and time dependent. There is no single recipe for effective risk communication.

Land-use planning and risk management

The United Kingdom, like some other European countries, has a system of control for major hazard installations, which recognizes the importance of the proper siting of facilities and the control of developments in the vicinity as a means of reducing residual off-site risk. The importance of planning controls has been reinforced in a 1996 European Directive (CEC, 1996). Siting control responsibilities (of new major hazards and of development in the vicinity) lay with the local planning authorities (LPAs), who also exercise control, as hazardous substance consent authorities, over the quantity and on-site location of hazardous storage. The local authorities received advice from the statutory regulatory agency, the HSE.

The control system has been operating for over 20 years. Over this period we have seen legislative activity to ensure that all potentially hazardous installations are identified and controlled (The Notification of Installations Handling Hazardous Substances Regulations 1982; The Control of Industrial Major Accident Hazards Regulations 1984; The Planning and Hazardous Substances Regulations 1992); the development of a structured consultation system between the LPAs and HSE over land-use planning proposals where the public may be at risk; and considerable advancement in the assessment of risks and presentation, or 'communication' of risk advice (Petts, 1988a; Petts, 1991; Miller and Fricker, 1993).

Siting decisions are made on a case by case basis, but utilizing general numerical risk criteria determined nationally. The adoption of this strategy reflects not only the needs of a decentralized control system, but also the information requirements of local decision makers and the variability in factors which influence risk decisions in different localities.

The use of the land-use planning system as a means of controlling the location of hazardous industry was endorsed by the Advisory Committee on Major Hazards (ACMH) precisely because it was seen as providing an opportunity for a local community to say whether, or not, it wanted a particular development (HSC, 1979). The same positive function of planning as a process of bargaining underpins views on the siting of potentially polluting industries such as waste incinerators. Whilst a facility may be able to meet the operational requirements of a regulatory agency, this does not necessarily mean that it is acceptable in land-use planning terms (Department of the Environment, 1994a).

The planning system provides a 'point of access' in a consensual system of regulation, for public lobbying of specific siting decisions. It is in the planning decision forum that many of the pressures on this consensual system - equity issues and demands for compensation, demands for freedom of information, the challenging of technical authority - are frequently highly evident. Planning is a mode of interest mediation (Healey et al., 1988), providing the forum in which conflicts over land-use, risk management and social priorities are first exposed, and the only direct forum in which they can be resolved. However, for the system to mediate effectively between interests it has to provide for participation. If arenas for public involvement are not provided, or information which is given to the public is limited and thus their ability to participate is limited, this produces problems. It has been suggested that people in the United Kingdom are more willing to accept risk decisions if the decision-making procedures are open and allow for informed public discussion of the results of a risk assessment (Sieghart, 1979). Fairness, or procedural accountability, will be determined not just by whether people are informed of proposals and asked for their views, but whether their views are genuinely taken on board so that they have an opportunity to directly influence decisions.

Public reactions

All of the psychometric studies identified in Chapter 12 seem to suggest that there is the potential for general public reaction against both the chemical and waste industry. In fact, in the United Kingdom at least, direct public reaction has been geographically erratic, small scale in terms of the numbers of people involved and localized (NIMBY: Not In My Back Yard).

Response to major hazards has, in the main, been temporally unsustained and with little involvement by the established environmental movement. A complex mix of social and economic factors appears to have an influence on perceptions, to the extent that, in some areas where hazardous industry has provided a significant economic benefit (such as Ellesmere Port, Cheshire), relatively high risks have been tolerated. Other mitigating factors in terms of perceived risk appear to be the degree of involvement of the industry in the local community and the record of the plants in terms of incidents and disturbance.

However, public reaction at the local level has certainly had an impact on the control system. For example, at Canvey Island, on the Thames, public concern over the agglomeration of petrochemical industry was instrumental in forcing a re-examination of siting decisions and consideration of the need to discontinue operations at a British Gas methane terminal (Petts, 1985). At Mossmorran (Scotland), a local action group forced a more detailed consideration of the risks involved in siting a new NGL plant and ethylene cracker (Macgill, 1982). In a few cases, operators have abandoned plans to locate at certain sites because of local concern, although some have been known to go elsewhere and to build with no, or very little, public reaction. It has to be said that experience of the siting of new major hazard installations on greenfield sites is comparatively rare now in the United Kingdom.

The Mossmorran action group fought a throughput expansion to the NGL plant in the late 1980s and the very capable and motivated local opposition has maintained its concerns over the years, particularly relating to their perceptions of the catastrophic or societal risks represented by the plant. It is interesting to note that despite a QRA now being available, the more objective and open discussion of risk levels does not seem to have diminished concerns amongst a small group of local 'opponents'. Certainly there were concerns that the QRA, in concentrating on individual risks, had not addressed the societal or catastrophic risks. However, there also appeared to be some inherent lack of trust in the company and indeed the HSE, owing to the fact that a now published study of the risks from ship/tanker movements had not previously been made public. There also appeared to have been concern that the HSE seemed happy to allow an expansion of the plant, but did not want an increase in the number of people at risk in the nearest community. Here we see a very distinct issue of public concern – a view that the statutory agency is applying double standards to risk decisions. Public perceptions of risk are compounded by a disagreement with the statutory agency as to how such risks should be managed, and there is suspicion that the 'experts' are working to some kind of hidden agenda (i.e., supporting industrial development to the detriment of a local community).

Waste facilities raise concerns about possible impacts which are common to many industrial developments (for example, traffic, odour, possible health impacts, safety, visual impact, loss of local amenity and decline in property values). In addition, however, examination of public responses (Petts, 1994; Wolsink, 1994; Petts, 1995) reveals a number of management concerns:

1. A lack of trust in a private sector industry to treat environmental protection as seriously as making a profit;
2. A lack of trust in either the operators or the regulatory authorities to monitor and control plant over the long-term;
3. A lack of trust in the state of knowledge about the health effects of emissions;
4. Concerns that the environmental and risk assessments do not adequately address the full nature and extent of potential impacts;
5. Questioning of the need for new facilities in the light of an apparent paucity of strategic planning.

Unlike the chemical industry, waste facilities rarely provide employment benefits and, in general, inequity in terms of local communities having to bear risks on behalf of others is a significant issue.

Certainly, from experience of local reactions in the United Kingdom to both major hazard installations and waste facilities, we have direct evidence of the limited relevance of psychometric approaches to the understanding of public risk perceptions and acceptability, and lessons for the importance of understanding the social and economic contexts in which risk decisions are taken if risk communication is to be successful.

Land-use planning decisions on risk – the influences

Public concerns can impact on planning decisions in that all local siting decisions are taken by committees directly accountable to the public –

local planning decisions are local community decisions. However, these local decision-making authorities operate under a whole set of constraints and pressures in their mediating role:

1.   *Planning precedent* – the precedent set by already well-developed urban areas where lack of awareness of hazard issues has led to inappropriate siting decisions in the past. Refusing further development can lead to concern about 'double standards' in terms of protection of an existing and a new population, as has been witnessed in relation to major hazards (Petts, 1988a; Miller and Fricker, 1993). Alternatively, existing sites (e.g. of waste incinerators) have been seen as preferred sites for new developments, despite a changed environmental setting and community issues (Petts, 1995).

2.   *Economic pressures* for development, which in certain parts of the country are very high. In relation to waste facilities there are pressures for development of incinerators in certain parts of the country with declining landfill capacity.

3.   *Land values* – if restrictions have to be placed on type of development which can take place (for example refusal of large scale housing development in the vicinity of a hazardous installation), this will have a direct impact on land values. Some authorities have expressed concern about potential 'urban blight'. Siting a potentially polluting facility in an area may have an impact on property values, although as yet there is no conclusive evidence on this in relation to waste facilities.

4.   *Lack of technical knowledge* – the majority of local planning officers and councillors are not technically trained in risk issues. In relation to major hazards they are dependent on the advice of the HSE. Increasingly, we have seen consultants playing a role in providing further independent advice to authorities particularly where an EA by the developer requires evaluation.

All land-use planning decisions relating to potentially hazardous activities involve consideration (direct and indirect) of a complex set of costs/risks and benefits (Petts, 1988b). For example, a proposed new major hazard activity may represent a new risk of harm to members of an existing population but may bring several hundred jobs to an ailing economy. A proposed large scale shopping and leisure complex in the vicinity of a major hazard plant may represent a beneficial use of derelict land and extend provision of local services. However, it will bring a new population into a risk zone and possibly restrict further development of the hazardous installation itself. An energy-from-waste incinerator in an urban area may reduce dependency on landfill and provide for resource recovery in line with sustainable environmental policies. However, it will bring increased traffic movements and potential air quality and health impacts. Different LPAs will respond in different ways to the pressures. For some, where economic or development pressures are high, certain risks may be tolerable. For others, the same risk levels could be deemed unacceptable.

Risk communication has to be able to deal with these complex costs/ risks and benefits. Experience in the United Kingdom has suggested that risk decisions relating to major hazards have largely (but not exclusively) followed the advice of the regulatory agency. For example, we understand that only a handful of authorities a year actually go against HSE advice not to allow specific development in the vicinity of hazardous installations (out of over 4000 such applications). Certainly the actual advice of the HSE relating to development in the vicinity of hazardous installations has developed in a pragmatic way seeking to accommodate planning pressures, e.g. certain low density land uses (small industrial developments, warehousing, etc.) are acceptable in the vicinity. However, such a 'success' rate if viewed from the statutory agency is not necessarily an indication of effective risk communication by the latter.

In relation to waste facility siting, some authorities have experienced adverse public reactions to required facilities and have had to revisit their waste strategies with inevitable delays. However, this is not a universal experience. The balancing of costs/risks and benefits varies locally, requiring different communication strategies. Effective risk communication must lead to an acceptable decision for the local community, a raising of awareness and understanding of hazard and risk issues, and justifiable trust in the capabilities of the controlling authorities. Concern amongst some authorities of the risk management strategy in relation to major hazards, and in particular its economic implications, has raised doubts about the effectiveness of risk communication. In 1990, the Association of County Councils criticized the HSE for seeking to control off-site risks through restrictions on land uses. The paper stressed that 'the reduction of the blighting effects of hazardous development should now be a major consideration for the government and environmental control agencies'. The paper recommended consideration of funding to allow local authorities to relocate firms away from sensitive populations or to pay firms to decrease risks to surrounding areas by reducing or removing amounts of hazardous materials stored on sites.

The HSE's risk criteria document (1989a) sought to discuss these concerns in an appendix which considered 'how to analyse net losses and gains to the nation from the introduction of the hazard and from taking explicit account of it'. It also comments on 'the transfer of wealth involved between the different parties'. The paper concluded that, *on a national basis*, the costs of planning restrictions measured against the benefits of greater safety are considerably less than the apparent effect on land values. However, this argument is unacceptable to planning authorities whose interests are only in the *local* costs and benefits.

Waste facility siting raises similar geographical dilemmas. The national waste strategy in the United Kingdom (Department of the Environment, 1995b) stresses overall targets for reducing waste and recovering resources from waste. It also supports the European 'proximity principle' which stresses that waste should be managed as close as is possible to its point of production. However, there may be tensions between this strategy at a national and regional level and its translation into local facilities. In the case of the Seal Sands hazardous waste incinerator (Petts and Eduljee, 1994) questioning of 'need' in the light of three different applications for incineration and a (claimed) out of date disposal plan forced a two-stage public inquiry to facilitate the local dispute. The risks from the plant had to be weighed against the regional need for a facility and a local perception that the County of Cleveland was becoming a dumping ground for other people's waste, with consequent stigmatization.

Risk information requirements for siting decisions

Risk communication pathways in siting decisions are quite complex, as detailed in Figure 15.2. The intensity and nature of risk communication varies according to the type of siting decision (e.g. local plan development or specific application for a new facility or development

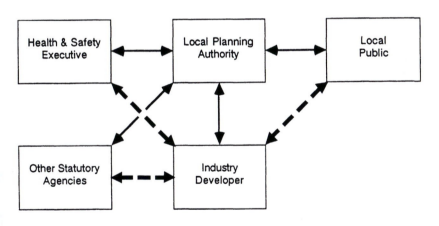

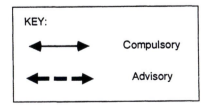

FIGURE 15.2
Risk communication pathways

in the vicinity), the stage of the siting decision (e.g. pre-formal planning application; planning application consideration; public inquiry) and also according to the knowledge/awareness levels of the local community and the inherent level of tolerance to risk activities. Whilst there are common information requirements in all of these situations, understanding of which is vitally important, it is also important to understand the specific requirements relative to the specific situation.

The information requirements in local decision-making reflect the decision-making constraints and pressures discussed earlier as well as the obvious need to consider the 'just how risky is this activity?' question. Examination of both major hazard and waste facility siting reveals that the information requirements include:

1. Information relating to the overall risk control strategy which the regulatory agencies (HSE or Environment Agency) support;
2. Information relating to the assessment of the hazards and risks. This includes the consequences for the full range of hazards, given the most likely and worst credible events (i.e. the ones the public fear), and the consequences both for the average and for the most vulnerable person. Also, information on the uncertainty in the assessment and the sensitivity of the results to this uncertainty. A QRA has become essential, most particularly where the risk decision is difficult and/or potentially controversial;
3. Information relating to facility operational control and monitoring;
4. Information relating to the balancing of the costs and benefits of risk mitigation options;
5. Information on the criteria against which the risks are judged.

There has been a significant development in the QRA information required in local decisions which has been most readily apparent in relation to waste facility risks in recent years. The focus is not upon the risk estimate per se (e.g. the individual's increased risk of cancer in a lifetime from exposure to emissions), but upon the assumptions and information which underlie the estimate. In siting decisions there has been evidence of the increased ability of people to access risk information from sources worldwide which is often used to challenge a proponent's, or regulatory authority's, assessment. This situation undoubtedly reflects the outstanding uncertainties which underpin assessments of chronic risks (in relation to acute risks, e.g. from fire or toxic releases, there is sometimes a greater degree of expert consensus). However, it also reflects the low level of trust in many experts and the increasing willingness and ability of people to challenge technical assessments.

Criteria of risk acceptability

The publication by the HSE of the criteria of acceptability (being revised at the time of writing), which it uses in relation to the land-use planning risk advice, was a step forward in improving the effectiveness of the risk communication process in relation to major hazards. The criteria are summarized in Table 15.2.

In the HSE document, risk is discussed in terms of receiving a 'dangerous dose' as opposed to the 'risk of death'. This criterion is more in sympathy with public fears of major accidents which cause suffering and disruption as well as death, and allows for better consideration of the more vulnerable individual than the risk of death to the 'average individual'. However, use of the criterion is not in line with other European practice which focuses on risk of death.

The criteria relate to specific types or categories of development. Thus, for small housing developments and retail, leisure and community facilities, a risk of 10×10^{-6} per year or lower of an individual receiving a dangerous dose was suggested as being 'acceptable'. For larger developments, a 1×10^{-6} figure and for those where particularly vulnerable people might be present, 1×10^{-6} to $\frac{1}{3} \times 10^{-6}$ was proposed. These criteria relate specifically to land-use characteristics and to perceptions of the need for extra protection for specific members of society. The approach depends upon effective categorization and the ability of the planning system to control the specific nature of development.

Criteria of acceptability around 1×10^{-6} are in line with other discussions of this subject. Discussions of chronic risks to health arising from waste facility emissions have followed international criteria adopted by the World Health Organization and United States Environmental Protection Agency of 1×10^{-6} increased lifetime risk. In 1984 the Department of the Environment set a target risk value of 1×10^{-6} per annum equivalent to a lifetime risk of 7×10^{-5}. The Royal Commission on Environmental Pollution (RCEP, 1993) agreed that a risk of 1×10^{-6} represents 'a reasonable upper bound'. However, it is interesting that in local decisions, where quantitative estimates have

Table 15.2 HSE criteria for development in the vicinity of major hazards

	Individual risk of receiving a dangerous dose (per annum)
Housing	10×10^{-6} to 1×10^{-6}
Retail community and leisure facilities	10×10^{-6} to 1×10^{-6}
Hospitals, schools, old people's homes	1×10^{-6} to 5×10^{-7}

been discussed, the HSE's major hazard criteria have usually provided a reference point.

The HSE's criteria only relate to individual risk. For societal risks, qualitative judgements are applied based on housing equivalence. There has been lack of a clear consensus on criteria for societal risk and difficulties in actually describing such a risk. Nevertheless, societal risk does equate with public concerns of catastrophic accidents and this is an area in which further progress in risk assessment and communication is required. In common with most discussions of major hazard risk, the HSE has favoured a comparative approach to risk communication. The risk criteria document includes a table of examples of other risks of death against which the major hazard land-use planning criteria may be compared (see Table 15.3). Unfortunately, in common with other such comparisons, the table shows risks from a number of 'voluntary' and occupational activities, exactly those types of activities for which public perceptions have been shown to be different from concerns over major hazards. Covello (1989; 1991) has summarized a number of the limitations and difficulties associated with risk comparisons which have been extensively discussed in the risk acceptability literature:

1. A failure to emphasize the uncertainties involved in the calculation of comparative risk estimates;
2. A failure to consider the broad set of quantitative consequences that define and measure risk, including a failure to provide risk data for sensitive or high risk groups; and a failure to include significant quantitative dimensions such as expected annual probability of injury or disability, spatial extent, persistence, recurrence, delay, expected environmental damage, etc. In the HSE's criteria which are based on risk of receiving a dangerous dose there is the obvious problem that comparative data that refer only to risk of death cannot be truly comparative;
3. A failure to consider the broad set of qualitative dimensions that underlie people's concerns about the acceptability of risks and technologies, discussed in Chapter 13;
4. A failure to consider alternatives to the technology or product being considered; to consider legal constraints to actions and the social consequences of risk decisions such as loss of privacy, the generation of social conflict or loss of civil liberties;
5. A tendency to draw on data from diverse sources that vary considerably in quality.

Whilst the risk communication literature has not discounted entirely the usefulness of risk comparisons, the choice of comparative data needs to take into account the concerns of the specific audience. Importantly, it is accepted that risk comparisons by themselves will not persuade anyone of the acceptability of a risk. An interesting study on

Table 15.3 Examples of individual risks (from HSE, 1989a)

Causes	Risk of death per million per year
All causes (mainly illnesses from natural causes)	11 900
Cancer	2800
(These figures vary greatly with age)	
All violent causes (accident, homicide, suicide, etc.)	396
Road accidents	100
Accident in private homes (average for occupants only)	93
Fire or flame (all types)	15
Drowning	6
Gas incident (fire, explosion or carbon monoxide poisoning)	1.8
Excessive cold	8
Lightning	0.1
Accidents at work – risks to employees	
Deep-sea fishing (UK vessels)	880
Coal extraction and manufacture of solid fuels	106
Construction	92
All manufacturing industry	23
Offices, shops, warehouses, etc. inspected by local authorities	4.5
Leisure – risks to participants during active years	
Rock climbing (assumes 200 hours climbing per year)	8000
Canoeing (assumes 200 hours per year)	2000
Hang-gliding (average participant)	1500

communication of risk information found that their use leads to greater opposition (Wright, 1993). Examination of siting debates relating to waste incinerators, where controversy and public focus on expert assumptions and risk assessment methodologies are important, suggests that the use of risk comparisons can be regarded as patronizing, reinforcing a lack of trust in the expert.

In a regulatory system where land-use planning is a locally exercised function then criteria of acceptability set at the national level are inappropriate. Criteria have to allow for local interpretation of acceptability relative to the economic and social needs and environmental priorities of a community at a specific time. There is direct evidence from planning decisions relating to major hazards that the acceptability of risk varies across authorities (Petts, 1988; Miller and Fricker, 1993). In waste incineration discussions, risk assessment has been used to provide a check on the site-specific acceptability of generic pollution control standards.

Costs/risks versus benefits

Whilst the quantitative assessment of risk and its evaluation using guideline criteria is a necessary tool of land-use planning decision-making, risk communication and information must also provide for discussion of alternative risk management strategies. In the United Kingdom the HSE's advice is based on a strategy to stabilize or reduce the population at risk from major hazards. Some planners express concern that in adopting this strategy the HSE is having to make implicit value judgements on land-use planning matters which are outside of its remit. Whilst the advice of an *independent* safety agency is important, the split of responsibilities between the HSE and the LPAs does lead to friction if there appears to be an overlap of interests. The open and public discussion of the control strategy is very important if such concerns are to be overcome.

On-site control of risk to a standard that is 'as safe as is reasonably practicable' involves a judgement on the part of the HSE on the costs of measures to avert risk and when such costs are in gross disproportion to the risk itself. This control is exercised without public discussion of the judgement involved, and is normally a matter of confidence between the industry concerned and the statutory agency. There is a need for the basis of such safety decisions to be openly discussed so that the costs of an off-site control strategy can be compared with the costs of the on-site control strategy.

The discussion of alternative risk management strategies in relation to waste has also underpinned communication problems. Opposition to proposals for new facilities is often centred on questions such as:

(a) Whether the facility is needed to deal with the waste arising in a local area;

(b) Whether the proposed facility is the right one for the types of waste, i.e. is it the best practicable environmental option (BPEO);

(c) Whether waste should be managed in some other way (Petts, 1992; 1994; 1995; Wolsink, 1994).

Whilst it is relevant to speculate that a focus on 'need' may in part stem from the perceptions of risk and disbenefit being an easier argument for objectors to prove than a case based on a complex assessment of health risks, it is nevertheless clear that perceptions of risk are not only based on physical consequences.

Management strategies for waste at the local level have traditionally been determined by the top-down approach with the private sector responding to general guidelines in the form of development plans which have suffered fragmentation and consultation deficiencies (Petts, 1995). There has been a tension between national and local consideration of the BPEO for particular wastes which, when translated into site-specific proposals, has seen public concern that fundamental risk–benefit decisions have either been taken with insufficient consultation or by people who will not ultimately face an application for development 'in their own back yard'. As with major hazard risk management, risk communication problems arise when important dimensions of the risk debate in the public context appear to have been discussed without opportunities for direct public input.

Another key area of discussion in the public domain is the effectiveness of risk control and reduction. For the public, understanding that a plant is operating as safely as is practicable will be achieved partly as a result of good operating record, thus allowing them to have some confidence in the control system. An open explanation and discussion of the mitigation measures that the installation is adopting may aid public acceptance. However, with increasing evidence that trust is important in risk perception, it has become apparent that members of local communities are looking for active involvement in the monitoring of operations (for example, through industry liaison groups), not just passive access to public registers of pollution incidents, emissions data, etc. Concerns about confidential information, the averaging of monitoring data over six-monthly periods, incident reporting, etc., are all evident in expressed scepticism of the value of public registers.

Effectiveness of the risk communication system

Both planning control over major hazards and the siting of new waste facilities suggest the need for effective risk communication. However, effectiveness in this context is seen to focus not only upon the development of assessment tools (particularly QRA) to improve confidence in the risk management process but also upon the means by which risk issues are discussed and communicated with interested parties. The latter will include the decision-makers (usually in the United Kingdom the local planning authority) and the public. However, the term 'the public' suggests a uniformity of group, interest, knowledge and concern that is rarely (if ever) evident in environmental/safety disputes (Petts, 1994). All individuals come to any issue with a plethora of interests and values and 'ways of doing things' (including risk assessment experts and government agencies). Risk communication has to face the challenge of new stochastic reasoning and at the same time deal with the fundamental conflicts between the perspective of the scientific/expert community and the public (Renn, 1992).

Risk communication must be based on the provision of information. This represents the base step on the ladder of public involvement (Arnstein, 1969), the position of least power and influence on decisions, but also the underpinning requirement for people to be able to take part in discussions. However, this chapter has illustrated that information provision from expert to decision-maker must be backed by other means, to enhance the opportunities for a consensus about risk decisions and management to be achieved. This is particularly important as all of the understanding of public responses indicates that the physical risks presented by a facility are not the only issue of concern and can rarely be isolated from other concerns about management competence and credibility, and related to equity in decision-making.

There is a need to find processes which facilitate involvement of all interested parties so as to provide for open disclosure and under-standing of different values and interests, while at the same time producing an informed judgement based upon a robust technical assessment. Problems arise where the statutory participative mechanisms and developed institutional ways of working are not amenable to adaptation. For example, the risk assessment process in relation to the siting of either a new major hazard plant or a new incinerator will involve informal discussions between the proponent and statutory authorities (HSE and environment agencies in the United Kingdom) during the development of the proposals. This early collaboration can positively assist in good project design. However, to a concerned community there may be suspicion of collusion between industry and the authorities and of agreement on issues without public involvement. The environmental assessment process provides opportunities for involvement of a range of interests in the early 'scoping' of what impacts to address, and there is considerable opportunity for this stage to include representatives of the public and local community as well as statutory interests.

Given the evidence of public concerns about fundamental cost/risk benefit decisions which are taken strategically, there is a need to address communication at the strategy and plan stages. A novel public communication programme instigated by Hampshire County Council in relation to the development of its strategy for managing municipal waste provides an example of the new approaches being tried in the face of opposition (Petts, 1995). The programme used communication approaches which sought to optimize opportunities for two-way discussion, particularly through small group discussions and seminars, and to provide for direct public influence on strategy development. The approach takes longer than traditional communication based upon passive and reactive consultation which require the formulation of plans and proposals prior to public involvement. The benefits may not be immediately visible. Nevertheless the Hampshire programme is indicative of new approaches to what has been termed 'rational discourse' which are being tested in different countries (Renn et al., 1995). In the general British planning arena, there are signs of willing-ness to adopt new community involvement tools (Department of the Environment, 1994b), but adoption is partial.

Effective communication has also to be considered in terms of the direct methods used and the skills of the communicator. The HSE has learnt that availability of officers to attend meetings to explain their risk advice, written detailed assessment, attendance at public meetings, etc. are all important communication tools. The skills of the risk communicator have been seen to be measured not only by an individual's technical ability and competence but also by the extent to which they show empathy and caring in relation to public concerns (Covello, 1992). When it is realized that trust and credibility are significant influences upon risk perception and public concern, then the approach of an individual and of their organization to communication becomes important. With the growing ability and willingness of interest groups to access information about industrial and environmental safety not only from official sources in their own country but also from sources worldwide, there is evidence that experts will be increasingly challenged on their assessments and decisions. Effective risk communication will require greater understanding of the concerns of different interests, the questions which are raised by these concerns and the sources of information which may be used to challenge experts.

Conclusions

Risk communication is a two-way process – a process of bargaining. Statutory authorities and industry must expect to learn and be prepared to change opinions and strategies, as must the public.

Risk communication which is perceived as simply risk education is unlikely to be effective, because it will almost certainly fail to address the main concerns of the public and information requirements of decision-makers.

Risk communication is an ongoing process – it is not simply a specific assessment in response to questions on a specific planning application. Discussion of risk control strategies; provision of information on site control and operations; on-going liaison between statutory authorities, industry, local communities and the media, all form an important part of risk communication.

Quantified risk assessment is now an essential element of siting decisions where risk is a dominant factor *or* is perceived by the public to be a dominant factor. Subjective discussion of 'small', 'low', 'insignificant', etc. risks is no longer acceptable.

Risk acceptability is location- and time-dependent. Risk communicators must understand the specific elements of the risk concerns for specific siting decisions.

References

Arnstein, S. (1969) Al Ladder of citizen participation, *J. Am. Inst. Planners*, 35, 216.
Baram, M. S. (1984) The Right to Know and the duty to disclose hazard

information', *Am. J. Pub. Health*, **74**, 385.

CEC (1996) Council Directive 96/82/EC on the control of major accident hazards involving dangerous substances, *Official Journal of the European Communities*, L 10/13.

Covello, V. T. (1989) Informing people about risks from chemicals, radiation, and other toxic substances: a review of obstacles to public understanding and effective risk communication, in *Prospects and Problems in Risk Communication*, W. Leiss (ed.), Institute for Risk Research, University of Waterloo Press, Canada.

Covello, V. T. (1991) Risk comparisons and risk communication: issues and problems in comparing health and environmental risks, in *Communicating Risks to the Public*, R. E. Kasperson and P. M. Stallen (eds), Kluwer Academic.

Covello, V. T. (1992) *Risk communication: a new and emerging area of communication research, in Risk Assessment*, Part 2, Proceedings of an HSE Conference on Risk Assessment, 6–9 October, London.

Covello, V. T. von Winterfeldt, D. and Slovic, P. (1986) Risk communication: a review of the literature, *Risk Abstracts*, **3**(4), 171.

Cvetkovich, G. and Wiedemann, P. M. (1988) Results of the working group: trust and credibility in risk communication, in *Themes and Tasks of Risk Communication*, H. Jungermann, Kasperson, and Wiedemann (eds), KFA, Julich, Germany.

Department of the Environment (1994a) *Planning and Pollution Control*, PPG 23, HMSO, London.

Department of the Environment (1994b) *Community Involvement in Planning and Development Processes*, HMSO, London.

Department of the Environment (1995a) *Guide to Risk Assessment and Risk Management for Environmental Protection*, HMSO, London.

Department of the Environment (1995b) *Making Waste Work: A Strategy for Sustainable Waste Management*, HMSO, London.

Fischhoff, B. (1995) Risk perception and communication unplugged: twenty years of the process, *Risk Analysis*, **15**(2), 137.

Gregory, R. (1989) Improving risk communication: a question of content and intent, in *Prospects and Problems Risk Communication*, W. Leiss (ed.), Institute for Risk Research, University of Waterloo Press, Canada.

Healey, P., McNamara, P., Elson, M. and Doak, A. (1988) *Land Use Planning and the Mediation of Urban Change*, Cambridge University Press, Cambridge.

Hirschhorn, J. S. (1984) Siting hazardous waste facilities, *Hazardous Waste*, **1**, 423.

HSC (1979) *Advisory Committee on Major Hazards*, Second Report, HMSO, London.

HSE (1989a) *Risk Criteria for Land use Planning the Vicinity of Major Industrial Hazards*, HMSO, London.

HSE (1989b) *Quantified Risk Assessment: Its Input to Decision-Making*, HMSO, London.

HSE (1992) *The Tolerability of Risk from Nuclear Power Stations*, HMSO, London.

HSE (1996) *Use of Risk Assessment in Government Departments*, HMSO, London.

Interdepartmental Working Group on Risk–Benefit Analysis (1984) *Risk–Benefit Analysis in the Management of Toxic Chemicals*, Agriculture Canada, Ottawa, Canada.

Kasperson, R. E. and Kasperson, J. X. (1984) *Determining the Acceptability of Risk: Ethical and Policy Measures*, Centre for Technology, Development and Environment, Clark University, Worcester, Mass.

Kasperson, R. E., Renn, O., Slovic, P., *et al.* (1988) The social amplification of risk: a conceptual framework, *Risk Analysis*, **8**, 177.

Lasswell, H. D. (1948) The structure and function of communication in society, in *The Communication of Ideas*, L. Brison (ed.), John Wiley, Chichester.

Layfield, Frank, Sir (1987) *Sizewell B Public Inquiry: Summary of Conclusions and Recommendations*, HMSO, London.

Mcgill, S. M. (1982) *Decision Case Study*, United Kingdom Mossmorran – Braefoot Bay 1982, CP-82-40, IIASA, Laxenburg, Austria

Marris, C., Langford, I. and O'Riordan, T. (1996) *Integrating Sociological and Psychological Approaches to Public Perceptions of Environmental Risks*, CSERGE Working paper GEC 96-07. Centre for Social and Economic Research on the Global Environment, University of East Anglia.

Miller, C. and Fricker, C. (1993) Planning and hazard, *Progress in Planning*, **40**(3), 171.

National Research Council (1983) *Risk Assessment in the Federal Government: Managing the Process*, National Academy Press, Washington, DC .

National Research Council (1989) *Improving Risk Communication*, National Academy Press, Washington, DC.

O'Riordan, T (1985) Approaches to regulation, in *Regulating Industrial Risks*, H. Otway and M. Peltu (eds), Butterworth-Heinemann, Oxford.

Otway, H and Wynne, B (1989) Risk communication: paradigm and paradox, *Risk Analysis*, **9**(2), 141.

Petts, J. (1985) *The Canvey Public Inquiries 1975-1982: Hazard and Risk Issues*, Working Paper MHC/85/1, Department of Chemical Engineering, Loughborough University of Technology.

Petts, J. (1988a) Planning and hazardous installation control, *Progress in Planning*, **29**(1), 1.

Petts, J. (1988b) Major hazard control: the local community and risk decisions, IChemE, *Symposium Series No. 110, 507*.

Petts, J. (1989) Planning and major hazard installations, in *Safety Cases*, F. Lees and M. L. Ang (eds), Butterworth-Heinemann, Oxford.

Petts, J. (1992) Incineration risk perceptions and public concerns: experience in the UK improving risk communication, *Waste Management and Research*, **10**, 169.

Petts. J. (1994) Effective waste management: understanding and dealing with public concerns, *Waste Management and Research*, **12**, 207.

Petts, J. (1995) Waste management strategy development: a case study of community involvement and consensus building in Hampshire, J. Env. Plan. Man., **38**(4), 519.

Petts, J. and Eduljee, G. (1994) *Environmental Impact Assessment for Waste Treatment and Disposal Facilities*, John Wiley, Chichester.

RCEP (1993) *Incineration of Waste*, 17th Report of the Royal Commission on Environmental Pollution, HMSO, London.

Renn, O. (1992) Risk communication: towards a rational discourse with the public, *J. Haz. Mat.*, **29**, 465.

Renn, O. and Levine, D. (1991) Credibility and trust in risk communication, in *Communicating Risks to the Public*, R. E. Kasperson and P. M. Stallen (eds), Kluwer Academic.

Renn, O., Webler, T., and Wiedemann, P. (eds) (1995) *Fairness and Competence in Citizen Participation: Evaluating Models for Environmental Discourse*, Kluwer Academic.

Royal Society (1983) *Risk Assessment, A Study Group Report*, London.

Royal Society (1992) *Risk: Analysis, Perception and Management*, London.

Sandman, P. M. (1988) Hazard versus outrage: a conceptual frame for describing public perception of risk, in *Themes and Tasks of Risk Communication*, H. Jungermann; R. E. Kasperson and P. Wiedemann (eds), KFA, Julich, Germany.

Sieghart, P (ed.) 1979 *The Big Public Inquiry*, Outer Circle Policy Unit, London.

Vertinsky, I. and Vertinsky, P.(1982) Communicating environmental health risk assessment and other risk information: analysis of strategies, in *Risk – A Seminar Series*, H. Kunreuther (ed.), IIASA, Laxenberg, Austria.

Watson, S., Fischhoff, B. and Hope, C (1984) Defining risk, *Policy Sciences*, **17**, 193.

Wolsink, M. (1994) Entanglement of interests and motives: assumptions behind the NIMBY theory on facility siting, *Urban Studies*, **31**(6), 851.

World Health Organization (1985) *Risk Management in Chemical Safety*, ICP/CE4, 506/m01, Geneva.

Wright, S. A. (1993) Citizen information levels and grassroots opposition to new hazardous waste sites: are NIMBYISTS informed? *Waste Management*, **13**, 253.

Wynne, B. (1989a) Frameworks of rationality in risk management: towards the testing of naive sociology, in *Environmental Threats*, J. Brown (ed.), Chapter 3, Belhaven Press.

Wynne, B. (1989b) Building public concern into risk management, in *Environmental Threats*, J. Brown (ed.), Ch. 8, Belhaven Press, London.

Wynne, B. (1991) Knowledge in context, *Science, Technology and Human Values*, **16**, 111.

Chapter 16

Safety management principles and practice

Introduction

The management systems and characteristics required to achieve success in health and safety are exactly the same as those required for success in any other area of industrial or commercial activity. Indeed, those organizations who are successful in business terms tend to be those who have high standards of health and safety at work.

Organizations are inherently complex. The adoption of new information and production technologies has tended to increase the diversity, if not the complexity, of the manager's task. In the area of safety, increased societal, social and technical pressures, together with legal constraints, demand the development of more efficient management systems and a better understanding of safety principles to guide their managers.

Safety has been defined by the International Standards Organization (Fido and Wood, 1989; Cox and Cox, 1996) as 'a state of freedom from unacceptable risks of personal harm'. This definition uses terms which are familiar to the reader. It also links safety with both risk assessment and with the elements of individual freedom. Safety management may thus be considered to be the management of this 'state'. In the context of systems thinking, 'safety management' is concerned with designing and maintaining reliable 'systems' which function in an expected and predictable (and thus 'safe') manner. It requires 'management' to make informed decisions in order to meet acceptable criteria. These decision-making processes are facilitated by the knowledge and understanding of systems reliability and of the potential hazards, risks and consequences discussed in earlier chapters.

Safety is also the focus of a plethora of international laws which reflect the policies of the countries in which organizations operate. Governments internationally acknowledge the need for systems of control and set standards through their respective 'safety' laws. In British courts 'safety' has been defined as the 'elimination of danger' (Latimer v. AEC Ltd, 1953), where danger is seen to encompass the probability of an accident and its potential consequences. Safety legislation may also provide a framework for managing safety, particularly in relation to recent changes in legislation philosophy. Regulations are increasingly based on risk management philosophies and the use of quantified risk assessment is often stimulated by legislative demands (see, for example, in the United Kingdom Control of Industrial Major Accident Hazards Regulations 1984 (CIMAH) and the Control of Substances Hazardous to Health Regulations 1994 (COSHH)). The Management of Health and Safety at Work Regulations 1992 discussed in Chapter 14 require specific arrangements for risk assessment within the United Kingdom. A review of legislation is outside the scope of this particular text. However, the interested reader can obtain further information on this topic from their appropriate enforcing authority, for example the United States Department of Labor, Occupational Safety and Health Administration and the Health and Safety Executive in the United Kingdom.

This chapter considers the implications of the principles of systems reliability and risk assessment for the management of safety. It highlights the importance of the management role in creating and supporting a safe working environment. It adopts a systems management approach to safety and outlines the key elements of safety programmes. It primarily emphasizes the need for clear objectives, policy and operational standards. Practical models of managing health and safety are discussed in the light of the first author's experience within a variety of organizations. An integrated approach to safety management is presented (see, for example, Cox and Cox, 1996) where safety, production and quality considerations share equal priority and the skills of 'good' management are the skills of safety management.

Systems management of safety: the principles

The systems approach to safety management is based on the application of general systems theory (see Chapter 1). It provides appropriate frameworks for analysing, modelling, and managing work and work functions, such as safety (see, for example, Singleton, 1984; Leplat, 1984; Groeneweg, 1994). There are several distinct stages in a systems approach to the management of safety.

1.   The determination of the safety system's overall function;
2.   The identification of the system's constituent parts and

functions and their interrelatedness, and the system's inputs and outputs;

3. The creation of a system's diagram or model (see Figure 16.1);
4. The evaluation of this model of the system using already solved problems or incidents (retrospectively) or against existing operations (concurrently);
5. Modification of the model;
6. Application of the model with monitoring and modification if necessary (predictive evaluation).

There are three important characteristics of 'systems' which can be applied to the simple safety systems model. First, the components are connected in an organized way and changes in one component affect the others. Second, the behaviour of the system changes if any one component is excluded from the system. Third the organized assembly of components does something. Safety systems also have an adaptive response to their wider environments (for example, the legislative environment, the geographical environment, the economic environment, etc.). They also operate in relation to systems goals, in this case the operation of a 'safe' system.

Safety programmes

In practice, such models translate into safety programmes characterized by a number of key processes and elements. Overall, these are concerned with formulating policy and objectives, organizing and planning to meet those objectives and putting in place and monitoring the necessary arrangements. Reviews of the leading organizations in relation to safety have identified the following key elements for success (Bird and Germain, 1987; HSE, 1989a; HSE, 1991; CBI, 1990):

1. Well-defined safety objectives;
2. Well-designed safety policy;
3. Demonstration of strong management commitment and competence (guidance, responsibility and accountability);
4. Adequate provision of resources for safety;
5. Agreed and clearly defined safety standards and procedures;
6. Joint consultation with the workforce;
7. Effective performance monitoring and feedback;
8. Effective incident investigation procedures;
9. Consideration of safety during selection and induction processes;
10. Systematic training programmes;
11. Promotion of principles of good job design in relation to safety: positive attitudes and (intrinsic) motivation, responsibility and meaning;
12. Effective communication with respect to safety;

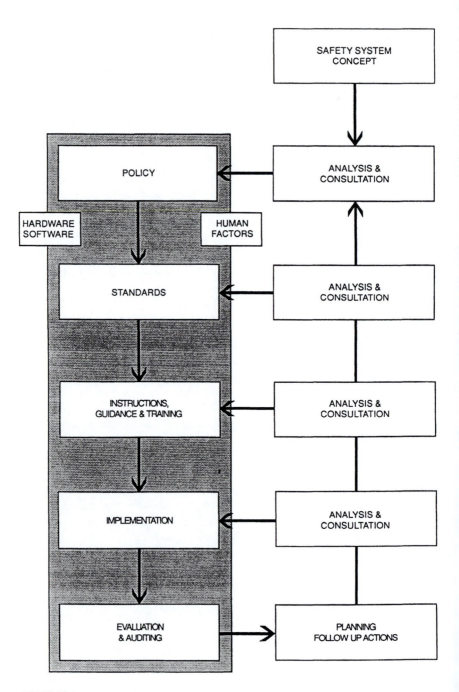

FIGURE 16.1
A systems model of safety

13. Well practised and effective emergency procedures;
14. The support of safety professionals.

Safety programmes can only be effectively implemented if safety is actively managed. Bryant (1984) highlights several examples of this process in organizations worldwide. He also stresses the importance of individual management qualities of leadership commitment and direction, together with clear management objectives, as the keys to successful implementation. 'Active' rather than 'reactive' approaches to safety management have been addressed by the first author in a separate text (Cox and Cox, 1996). However, some of the key elements and processes of safety management are further elaborated in this chapter.

Safety goals and objectives

Figure 16.2 presents an 'objective setting' model for the 'safety' of System 'X'. It represents the sequence of operations involved in the design and operation of safe systems. Implicit in this model is the requirement to analyse the system, to set objectives and standards in line with corporate policy, to create the organization and arrangements necessary to meet these objectives and to establish meaningful methods of monitoring. Information (gained through the process of risk assessment) is the 'life blood' of this model and 'decision makers' have the opportunity to bring it to 'life'.

How then should the analysis proceed, how may the necessary information be presented and how can the management (decision makers) utilize it in the design of safe systems?

Systems analysis

There are several techniques available to the 'systems analyst'. These have been variously described (Checkland, 1981; Oliga, 1988; Waring, 1989). Every technique requires a description of the system as a starting point. In a safety context this description not only includes the active components of the system but it also identifies the associated hazards and risks. This preliminary description is usually presented diagrammatically. Three types of diagram are commonly used in systems analyses as an aid to thinking:

1. Organization charts, systems maps and influence diagrams (structures and relationships);
2. Flow charts, decision sequence, flow block and data flow (process charts);
3. Rich pictures and spray diagrams (thinking aids).

The main conventions or diagramming rules are Venn and digraph (Waring, 1990). Figure 16.3 presents a spray diagram for a process involving the use of lead (see Checkland, 1981, for examples of the other diagrams).

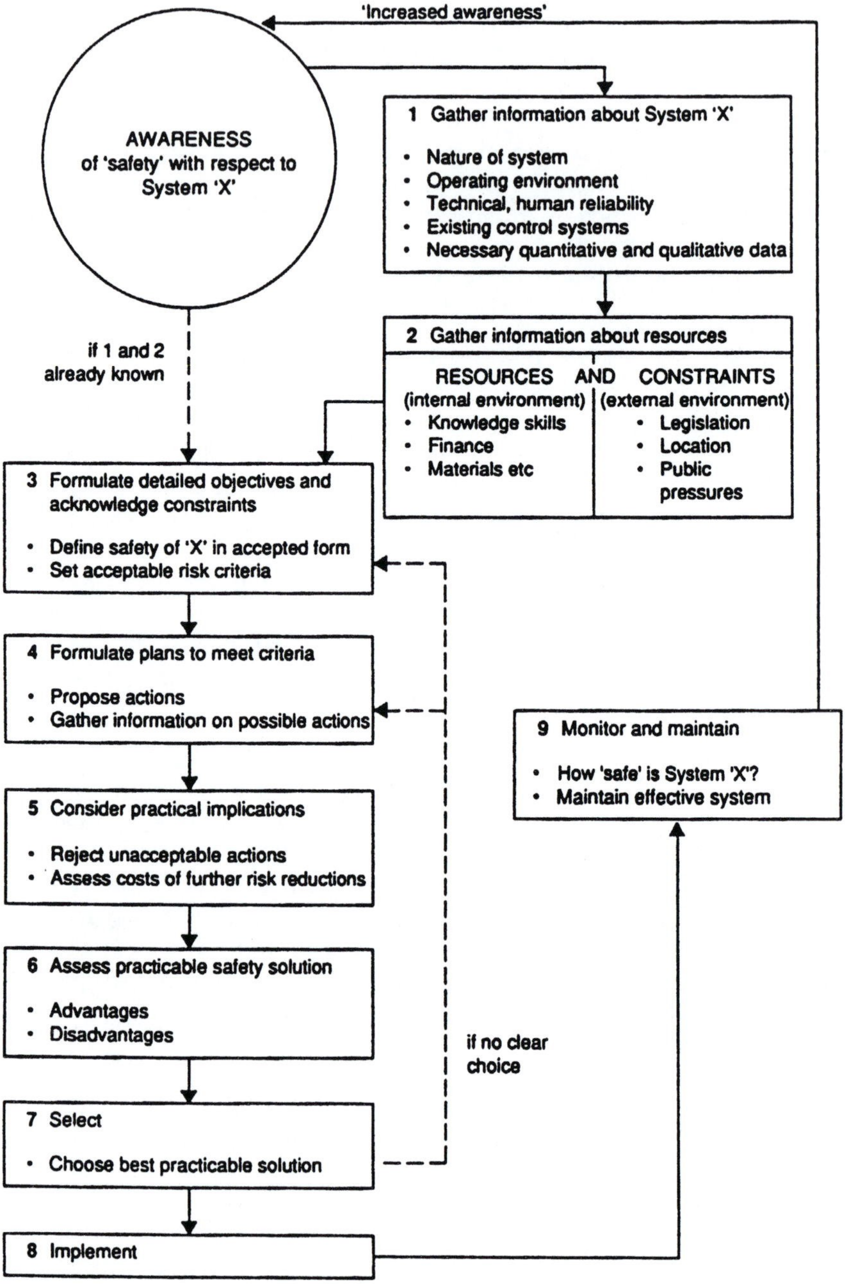

'Increased awareness'

AWARENESS
of 'safety' with respect to
System 'X'

1 Gather information about System 'X'

- Nature of system
- Operating environment
- Technical, human reliability
- Existing control systems
- Necessary quantitative and qualitative data

if 1 and 2
already known

2 Gather information about resources

RESOURCES AND CONSTRAINTS	
(internal environment)	(external environment)
• Knowledge skills	• Legislation
• Finance	• Location
• Materials etc	• Public pressures

3 Formulate detailed objectives and acknowledge constraints

- Define safety of 'X' in accepted form
- Set acceptable risk criteria

4 Formulate plans to meet criteria

- Propose actions
- Gather information on possible actions

9 Monitor and maintain

- How 'safe' is System 'X'?
- Maintain effective system

5 Consider practical implications

- Reject unacceptable actions
- Assess costs of further risk reductions

6 Assess practicable safety solution

- Advantages
- Disadvantages

if no clear
choice

7 Select

- Choose best practicable solution

8 Implement

FIGURE 16.2
An 'objective' setting model for the 'safety' of System X

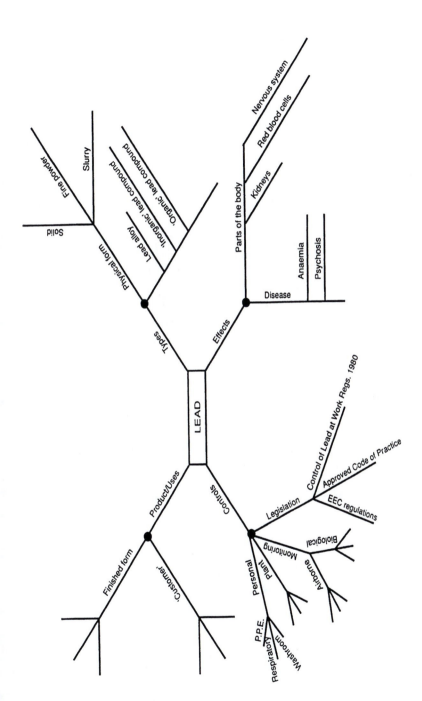

FIGURE 16.3
Spray diagram for a process involving the use of lead

The next stage requires the analyst to define the purpose of the system. Using the 'lead' example, it may be the formulation of a 'safe' system for handling lead as part of the overall safety policy and objectives. They then consider the characteristics of the system required to meet such objectives and define acceptable performance criteria. At this point they may need to refer to organizational safety policies.

Safety policies, standards and procedures

A safety policy sets out the organizational goals, responsibilities and arrangements for 'safety'. In some countries (for example, within the United Kingdom) there is a legal requirement to produce a written statement of policy (Health and Safety at Work, etc. Act 1974). It should reflect the interest organizations have towards health and safety and clearly indicate that such commitment comes from the 'top'. This commitment is usually expressed in a statement of guiding principles. For example, company Y has a policy statement which formulates the following objectives (CBI, 1990): 'It is the policy of company Y to practice and maintain a safe and healthy working environment for all its employees, to review continuously all practices and procedures which could affect their health, safety and welfare at work and to ensure that any necessary improvements are implemented immediately'.

An effective policy will not only help to reflect management commitment to safety but it will also support the control of organizational operations (e.g. safe handling of 'lead'). In risk management terms, the policy should be concerned with establishing operational limits and standards within 'acceptable' risk parameters and making suggestions for the reduction of risks (Fido and Wood, 1989). In order to progress this requirement, management needs to acquire and utilize the necessary information. The spray diagram (see Figure 16.3) focuses attention on specific information 'needs'. This information should be presented in a form which is readily understood and which is consistent with the operational environment in which standards have to be set and performance parameters should be clearly defined (HSE, 1991).

Quantified measures of system performance require consideration of reliability data for various components of the system. These data can be obtained from reliability and risk analyses (see earlier chapters) and can be stated in measurement terms (for example, clear and concise occupational exposure standards). However, additional quantifiable measures of system performance need to be defined (for example, compliance costs) and monitoring techniques assessed. The performance monitoring of systems for 'safety' will be discussed in a later section. Finally, an assessment of risks also provides the analyst with the necessary management information to discuss 'control' and to formulate an effective and safe system (i.e. to manage the risks).

Decision making and the design of safe systems

The decision sequence diagram enables managers to 'assess' and 'manage' operational risks (Figure 16.2) through the design process. It also makes reference to the wider system environment (see stage 2). This includes social processes and legislative guidelines and standards. Such legislation is not only aimed at protecting the immediate users but also the public at large (see Chapter 15). It forms the basis of acceptable standards. Social and humanitarian consider- ations should include the general well-being of employees and interactions with the public who either live near the organization's premises or who come into contact with the organization's operations or products (Raafat, 1989). Perceived risks (see Chapter 13) and individ- ual and public acceptance should also be taken into account. This is a particularly difficult area for managers to accommodate and guidelines on managing subjective risk are available from external consultants and industry and trade organizations (Lowrance, 1990). Decisions should also encompass the possible benefits generated by the particular operation together with both political and economic considerations.

Costing accidents

Economic considerations include not only insured costs but also the uninsured costs of accidents or systems failures. Various estimates of these costs have been made (see Chapter 11) and these have exploited a variety of costing techniques. Research into accident costs carried out on behalf of the International Labour Organization (Andreoni, 1986) has identified several elements associated with safety related expenditure. These include:

1. Routine expenditure incurred before occupational injuries arise (preventive expenditure);
2. Expenditure following the occurrence of occupational injury;
3. Expenditure representing transfer to an insurance organization of some of the financial consequences of occupational injuries;
4. Exceptional expenditure on prevention.

Part of these expenses is fixed; more precisely, it can be considered as practically invariable, being related to the events that accompany the production operations. Total expenditure may be expressed mathematically as:

$$D_d = D_{pf} + D_{af} + D_{pv} + D_{av} + D_1 + D_m + D_{pe}$$

where the cost elements are:

D_d = total expenditure during the course of production
D_{pf} = fixed expenditure on prevention

D_{af} =   fixed expenditure on occupational injury insurance
D_{pv} =   variable expenditure on prevention
D_{av} =   variable expenditure on occupational injury insurance
D_{1} =   variable expenditure resulting from occupational injuries
D_{m} =   variable expenditure arising from material damage related
               to occupational injuries
D_{pe} =   prevention expenditure of an exceptional nature.

These elements may be defined as (Andreoni, 1986):

(D_{pf}) – *Fixed expenditure on prevention* includes the total expenditure on the safety and health services, safety representative and committee costs, expenditure related to occupational hygiene, etc. and administrative costs on records. In theory all these types of expenditure are necessary. In practice, they depend on the provisions of regulations governing the operation of each individual organization.

(D_{af}) – *Fixed occupational injury insurance expenditure* which is of several types. For material danger there are special types of insurance (fire, industrial, subsequent losses, etc.). In the field of occupational injuries the costs are essentially those of insurance against such injuries. Premiums vary from country to country as does their mode of calculation.

(D_{pv}) – *Variable expenditure on prevention.* This is dependent on the frequency and severity of incidents and additional training aims incurred, research studies, additional propaganda, etc.

(D_{av}) – *Variable expenditure on occupational injury insurance* depends above all on the insurance arrangements, which can vary from one country to another or from one sector of activity to another.

The private (mutual benefit) insurance schemes and some social security schemes fix the level of the premiums either on the basis of the occupational injuries which have occurred in the enterprise or on the basis of its potential degree of risk. Premium reductions or increases can take place subsequently (depending on the provisions of the insurance policy) in the light of changes that occur in the frequency and severity of occupational injuries that have occurred during a given period. It can happen, for example, in the case of an inexcusable fault by the employer, that the enterprise is made to repay the sum paid to the victim by the insuring institution; these cases are rare but costly.

(D_{1}) – *Variable expenditure on occupational injuries.* This includes two basic categories of expenditure:

1. First aid given on site, transport costs, external medical treatment, legal costs, fines, etc.
2. Wages paid during the period of absence and wages paid to other workers who may be inactive at the time of the accident.

(D_{m}) – *Variable expenditure arising from damage linked with occupational injury* – varies according to the nature of the incident.

(D_{pe}) – *Exceptional expenditure on prevention* – outside the fixed nature prevention (e.g. additional protection against noise).

It is important for management to consider all the elements in the risk assessment. If the fixed expenditure on 'prevention' is exceedingly high and the risk consequences are low then some adjustments may be made. However, if the consequences including expenditure of systems failure are high this may be a totally justified cost.

Cumulative costs of accidents, ill health and incidents of property damage are also an important feature of organizational costs. Many organizations are unaware of such costs as they are often 'hidden'. The obvious costs are those arising from damage (D_m), legal costs and repairs (part of D_l) and increased insurance premiums (D_{av}). But each accident, incident of ill health or damage to property also results in lost production through employee absence, through plant and equipment being out of use and involvement of staff in investigation and restoration. This may also lead to a loss of business and goodwill. Organizations may also lose valuable skills and require expensive retraining. All this costed out can add up to a considerable wastage in human and material terms.

One organization summarized these losses as follows (CBI, 1990): 'Although accident costs are largely met by insurers it should be appreciated that each accident is also a direct cost to the company. It is estimated that every accident, whether investigated or not, involves a minimum non-recoverable cost of some £1500 which, for the accidents sustained this year, comes to at least £141 000. As almost all accidents are avoidable this is a totally unnecessary waste of money'.

In summary, the assessment of risk involves not only the consideration of the probability or frequency of each potential 'system failure' and the severity of the outcome, it also involves economic, social, political and legal considerations. The magnitude of the loss is balanced against the possibility of control. It is important to note that the prevailing philosophy within the UK towards decision making for safety demands that risk analysis is iterative (HSE, 1989b; 1991). That is to say that further reduction of risk should repeatedly be attempted taking reasonable costs into account.

Risk control
Risk control strategies may be classified into four main areas: risk avoidance, risk retention, risk transfer and risk reduction. These elements are described in detail in Bird and Germain (1987) and in Crockford (1980). Risk avoidance is a strategy in which an organization consciously decides to avoid completely a particular risk (for example, an organization may decide that the risks associated with handling a particular substance are too great and cease manufacture).

There are two aspects of risk retention: with knowledge and without knowledge.

1. With knowledge – a conscious decision is made to retain (via self-financing) the risk within the organization. This may involve the organization in the practice of negotiating an excess on their insurance premium.
2. Without knowledge – this occurs if all the risks have not been evaluated, and thus if an 'occurrence' has an effect on the organization it has to be paid for by that organization.

Risk transfer refers to the legal assignment of the costs of certain potential losses from one party to another. The most common way of effecting such transfer is via insurance.

Risk reduction, often termed loss control, considers the reduction of risk within the organization by the implementation of a loss control programme. The basic aim is to protect the organizational assets from wastage caused by accidents or systems failures. Initially there is a need to collect data on as many loss-producing incidents as possible in order to set up a programme of remedial action. The elements of a loss prevention programme are integrated into our safety systems model in Figure 16.4.

Performance monitoring and evaluation

Performance monitoring and evaluation is a fundamental part of risk control. Measurement of performance involves comparisons with standards. Without adequate standards there can be no meaningful measure of performance (Bird and Germain, 1987). The previous sections have discussed 'standard' setting against acceptable criteria and have outlined the elements (or subsystems) of risk control programmes for all work activities (see Figure 16.4). It is important to establish clear and demanding standards for all the elements listed in 'Box A' and for all work activities. In other words, performance criteria should be established for plant, personnel, management activities such as selection and training and for the effectiveness of the overall system (Petersen, 1971). For example, the British Standard 8800 Guide to Safety Management Systems (BSI, 1996) lays down standards for safety management systems. The systems described within the BSI document provide a framework for action which reinforces the approach taken by the United Kingdom HSE (HSE, 1991). Performance monitoring is thus not restricted to the recording of accidents and incidents (as discussed in Chapter 11) which are the traditional parameters of 'safety', but it is also an 'active' process. This philosophy of safety measurement is concerned with actively seeking information on systems with a view to promoting positive actions (see earlier sections). It provides some support for a quote from Peters and Watermann (1982): 'what gets measured gets done. Putting a measure on something is

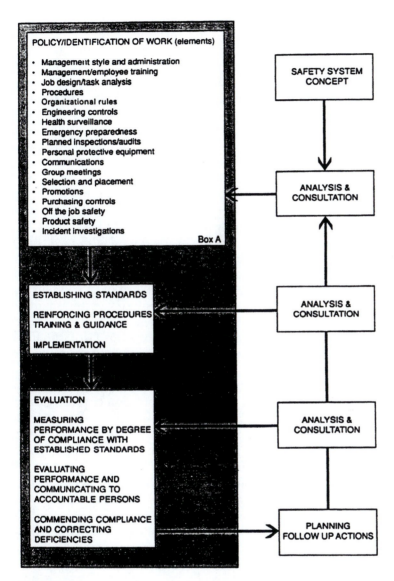

POLICY/IDENTIFICATION OF WORK (elements)

- Management style and administration
- Management/employee training
- Job design/task analysis
- Procedures
- Organizational rules
- Engineering controls
- Health surveillance
- Emergency preparedness
- Planned inspections/audits
- Personal protective equipment
- Communications
- Group meetings
- Selection and placement
- Promotions
- Purchasing controls
- Off the job safety
- Product safety
- Incident investigations

Box A

SAFETY SYSTEM CONCEPT

ANALYSIS & CONSULTATION

ESTABLISHING STANDARDS

REINFORCING PROCEDURES
TRAINING & GUIDANCE

IMPLEMENTATION

ANALYSIS & CONSULTATION

EVALUATION

MEASURING
PERFORMANCE BY DEGREE
OF COMPLIANCE WITH
ESTABLISHED STANDARDS

EVALUATING
PERFORMANCE AND
COMMUNICATING TO
ACCOUNTABLE PERSONS

COMMENDING COMPLIANCE
AND CORRECTING
DEFICIENCIES

ANALYSIS & CONSULTATION

PLANNING
FOLLOW UP ACTIONS

FIGURE 16.4
Loss prevention elements: a systems model

tantamount to getting it done. It focuses management attention on that area. Information is simply made available and people respond to it'.

It helps to focus on the success of control measures for 'safety' rather than the occurrence of 'failures'. Such an approach therefore needs managers to fully understand the reliability of the whole system. It also requires the necessary ongoing checks of system behaviour discussed in earlier chapters.

In practice, organizations use audits of policy, procedures, plant and people to maintain consistency and to provide valuable performance indicators (HSE, 1985; 1991). Safety audits provide a method of examining actual performance against the yardstick set in the standards and work systems set down in the safety policy and safety manual. Auditing can take place at various levels of an organization from the level of the enterprise down to individual facilities and plant. The auditing process should involve all employees and auditors should ask the question 'what do you think about the safety around here?'. There are a number of proprietary audit systems on the market including the International Safety Rating System (ISRS) (Bird and Germain, 1987), Complete Health and Safety Evaluation (CHASE, 1987) and RoSPA Five Star audit (RoSPA, 1996). There are also specialized audit schedules for specific industries (for example, the chemical process industry). An important aspect of any audit system is its ability not only to measure what is in place but to gauge actual compliance with required performance standards (BS 7229, 1989).

Accident and incident investigation is also regarded as an important part of monitoring 'safety'. An understanding of the chain of circumstances and trigger events (systems failures) that lead to an accident can highlight 'systems' faults. A methodology for analysing systems failures is described in Waring, 1989. It contains two case studies of relevance to health and safety. One concerns the failure of a hoist in Littlebrook Power Station (1978) and the other concerns structural failures in housing developments.

In order for accident investigations to be effective they should involve senior management and be carried out as soon as possible after the incident (CBI, 1990). The outcomes of these investigations should also be fed back to all members of the organization to support further learning. Some organizations incorporate 'safety' performance into annual appraisal of employee performance. This is just one example of how safety can be integrated into mainstream management activities. Other examples incorporate the introduction of health and safety training into supervisor or team leader development courses (Cox and Cox, 1996), the incorporation of safety suggestions into innovation awards, and the incorporation of health and safety performance into annual reports. The prevailing philosophy is one of integration. It is based on the premise that organizations who manage safety well are usually effective in other domains (Dawson, 1997). The 'practice' rather than the 'principles' of systems management of safety are outlined in the following section.

Systems management of safety: the practice

Management has responsibility for ensuring that safety policies are translated into working practice. They thus have to be communicated, implemented and maintained at all levels of the organization. For example, organizational structures need to be established which facilitate rather than work against the policy. Clear lines of responsibility should be defined and all personnel should understand their role and fully participate in the process (HSE, 1989a; 1991). Well-defined safety standards and procedures should be in place for all operations (see earlier), including maintenance operations, and these should be monitored on a regular basis. Training programmes should be implemented wherever possible in accordance with good training practice (Cox, 1991a) and all employers should be encouraged and motivated to make safe systems work.

The authors have stressed the importance of an integrated approach to safety management (Cox and Cox, 1996) and all the programme elements should be integrated into general management systems whenever possible. For example, 'total quality' management systems are based on four principles of quality (Crosby, 1979; Juran, 1988):

1. *Definition of quality* – quality is conformance to agreed and fully understood requirements, meeting what the customer wants in every way. There is no high or low quality. The product or service either satisfies what the customer wants or it does not;
2. *System of quality* – prevention rather than appraisal. It involves focusing on work processes rather than on people to reveal where deviations occur, and then taking action to prevent them. It can be regarded as the application of good planning techniques to every action taken in an organization;
3. *Standard of quality* – zero defects. This means meeting the customer's requirements without defects;
4. *Measurement of quality* – by pricing non-conformance. By calculating the cost of doing something wrong, it is possible to understand what the financial impact on the business is.

These principles could be easily applied to 'safety' (Whiston and Eddershaw, 1988 and Fido and Wood, 1989). Although organizations have adopted total quality management and integrated it into the corporate culture at all levels, many still see safety management as an afterthought. The elements for a quality programme are those of planning, controlling, inspection, corrective action, communication and review and they all require working procedures. This review is analogous to the safety systems 'audits' described earlier and may be carried out by external and internal auditors. In practice, a large percentage of companies in the United Kingdom are audited by a third party to ensure compliance with the British Standard (BS5750, 1987) or International Standard (ISO 9000/1) on quality systems. This is analogous to the systems approach to safety discussed earlier and there is a practical benefit in linking the two programmes to develop a 'safety' and a 'quality' culture.

Safety culture

Management specialists (Likert, 1967) have long suggested a link between organizational culture and the success of the organization. It is often described as the mixture of shared values, attitudes and patterns of behaviour that give an organization its character (ACSNI, 1993). A positive safety culture is therefore reflected not only in the general state of the premises and conditions of the machinery but in the attitudes and behaviours of the employees towards safety (see later). Some organizations have developed such a culture of health and safety. It is an integral part of their business. Others believe that safety cultures can be developed in isolation of business considerations or that safety is too costly and time consuming.

The CBI (1990) carried out a survey of 400 firms and examined the safety performance of another 50 companies to explore ways of developing safety culture. Table 16.1 summarizes the essential features of a sound safety culture identified in this survey.

Many of the organizations surveyed reported the need for active employee participation in solving safety problems, in formulating safe working procedures and in developing safety culture. They observed that documented standards could only work in practice if the people involved in the task carried out the procedures in an interested manner. There are a number of ways in which organizations can encourage employee participation for safety. These include involving employees

Table 16.1 Essential features of a developing safety culture (CBI, 1990)

Leadership and commitment from the top which is both genuine and visible

Acceptance of long-term strategies and sustained effort and interest

A policy statement of high expectations which conveys a sense of optimism about what is possible

Adequate codes of practice and safety standards

Health and safety treated as seriously as other corporate aims, and properly resourced

Line management responsibility for health and safety

'Ownership' of health and safety permeating all levels of the workforce (this requires employee involvement, training and communication)

Realistic and achievable targets and performance standards

Audits of the whole 'system'

Incident investigations

Consistency of behaviour against agreed standards with good safety behaviour as a condition of employment

Deficiencies revealed by an investigation or audit should be remedied promptly

Adequate and up-to-date information to enable management to assess performance in committees and safety projects, training, communicating safety and implementing motivational schemes.

Joint consultation and safety committees

Employee involvement in health and safety is a feature of working life in most industrialized countries. Trade union appointed safety representatives and safety committees (when requested) are a legal requirement in the United Kingdom (Safety Representatives and Safety Committees Regulations 1977). The European Commission's 'frame-work' Directive (1988) also requires member states to establish workers' representatives and to encourage employee proposals and employee representation. This has been incorporated into the recent United Kingdom regulations on employee consultation (HSC, 1996). Organizational safety committees are often seen as a practical and convenient way of fostering such involvement in health and safety within organizations. However, it is not sufficient just to establish a committee; it has to play an active role within the organization. Committees have to agree terms of reference which set out their objectives, constitution (membership) and frequency of their meetings. A review on employee participation within the United Kingdom (Walters and Gourlay, 1990) provides data on the presence, activities and training of safety representatives together with the existence, role and composition of safety committees. It also highlights a number of weaknesses in the operation of the 1977 Regulations which have now been addressed in the 1996 regulations. These include lack of employee involvement in small firms, declining level of safety representative training, inadequate trade union organization and employer support.

Safety committees and employee representation also operate successfully in non-unionized companies outside the Safety Representatives and Safety Committees Regulations (1977) alongside less formal methods of employee involvement. These schemes include employee suggestion schemes, small group 'toolbox' talks and problem solving teams (sometimes called 'safety' circles rather than 'quality' circles). An alternative to small groups of employees in similar employment is a 'diagonal slice' group in which employees in different functions and levels come together to solve safety-related problems.

Safety education and training

Safety education and training have become key strands in many accident prevention and safety promotion strategies within organizations (Dawson et al., 1988). However, it is questionable whether the quality of such education and training programmes has been sufficiently high to produce the expected (and required) return. This is largely due to the failure to address properly all the different processes which underpin education and training and which make up the training cycle (Cox, 1988a). Hale (1984), in a review of safety training, highlighted the fact that most general reviews on this subject (for example, Hale and Hale, 1972) begin their discussions with the comment that remarkably few studies have been published which evaluate the effectiveness of safety training. When questioned about safety education and training, many managers immediately focus on the actual training session; however, this is but one part of an overall training cycle, which begins with the training needs analysis and culminates in the evaluation of training (Cox, 1988a).

Safety education and training normally seek to fulfil two linked objectives. First, they attempt to improve individuals' awareness, knowledge, attitudes and skills in relation to health and safety, as reflected in safe working behaviour and procedures. Second, they also attempt to effect positive changes at the level of the organization itself. These include an obvious improvement in task performance and in overall safety performance and an effect on 'safety culture'. Both objectives are derived (in detail) from the training needs analysis and must be reflected in the later evaluation of training.

Training needs analysis

Logically the training cycle begins with the systematic identification of training needs achieved through a training needs analysis. Among other things, this analysis must consider the needs of the organization in relation to the different environments in which it has to operate. The training needs analysis should be designed to consider each working group within the organization in relation to its safety behaviour and to the organization's overall safety performance. For example, in the chemical industry (Cox, 1991a) these groups may include:

- Safety professionals
- Plant process operatives
- Instrument artificers and technicians
- Craftsmen and fitters
- Engineers
- Maintenance personnel
- Supervisors and plant management
- Sales personnel
- Contractors.

In each case, a task analysis should be completed and safety relevant elements of the task identified along with associated safe behaviours (Bamber, 1983; Stranks, 1990).

This sort of analysis should allow shortfalls in safety awareness and knowledge, attitudes, skills and workplace behaviour to be identified, group by group, along with priority groups for receiving education and training. By the same process, priority issues or hazards can be identified, group by group, and at the level of the whole organization or even industry.

A questionnaire-based survey by Sandra Dawson and her colleagues at Imperial College, London, on safety in the chemical industry (Dawson *et al.*, 1988), asked respondents (production managers, supervisors and engineering maintenance personnel) to identify: 'the most important health and safety problem or hazard faced by the people for whom they (managers or supervisors) were responsible or whom they (safety representatives) represented'. Their results are set out in Table 16.2. These data may be used to support data derived from the task analysis. Interestingly, the data results demonstrated a relatively high level of awareness within the United Kingdom chemical industry of issues relating to human factors. They also pointed out the need to be particularly alert to the needs of several 'at risk' sub groups: new employees, transferred employees, temporary employees, and promoted employees.

Detailed objectives for education and training are developed from the analysis of training needs, possibly guided by a 'framework' model of safety behaviour and its management. Not only does the publication of those objectives allow trainees to understand what that training is about, they also effectively guide the rest of the training cycle and focus on necessary job competencies. The essential question is: 'What should trainees be able to do at the conclusion of training, and how will this differ from what they can do before they are trained?'

Objectives should thus be stated in behavioural terms for the benefit both of the trainees and of later evaluation. Often this presents terminological and conceptual difficulties when training is targeted on knowledge, attitudes or attitude change (Cox, 1988a). Such difficulties can be overcome with some imagination. What, therefore, is the best way of increasing trainees' awareness, knowledge, attitudes and skills (AKAS) to allow (enable) them to meet the behavioural objectives set?

Decisions on appropriate training methods will be dependent on a number of criteria (Cox, 1991a), including:

1. The nature of the subject matter;
2. Learning objectives (AKAS);
3. The number of trainees;

Table 16.2 Respondents' ranking of health and safety problems (Dawson *et al.*, 1988)

Nature of hazard	% respondents
Chemicals (toxicity, corrosive, fire, explosion)	37
People (attitudes, carelessness, ignorance)	29
Plant, machinery and systems of work	17
Place of work	12
Other	4

4. Trainee preferences and prejudices;
5. The total training resource;
6. The convenience factors such as shift patterns and job cover.

Economies of scale are provided by a large number of trainees attending a lecture but better retention of information occurs in more participative scenarios, including role plays. At the same time, computer-based training methods allow greater flexibility for individuals within organizations and make allowance for differing rates of learning.

Training materials are available from a variety of sources to support safety training and important technological advances have been made in this area with the advent of video discs. Similar developments in computer software (e.g. customized permit to work packages) have standardized systems information. Institutions such as the Institution of Chemical Engineers in the United Kingdom and the American Institute of Chemical Engineers have produced training packages on a number of safety-related topics including 'emergency preparedness' (Cox, 1991b). Distance learning materials are also available to industry, often prepared in conjunction with educational establishments such as the Centre for Hazard and Risk Management (Loughborough University). It is important to determine whether education and training have been successful. Training programmes can be evaluated in different ways:

1. By examining trainees' progress during the course and their reactions to it at the end of the course (internal evaluation);
2. By examining the impact of the training on the trainee's later job performance, etc. (external validation of key competence);
3. By asking whether the training has achieved its organizational objectives.

Some computer-based training packages build in not only an examination of trainee achievement against programme objectives, but also map their progress in doing so (internal evaluation). Evaluation needs to be carefully planned into the training cycle, with the design of the evaluation beginning during the training needs and organizational analysis.

Communicating safety

Communication is a vital aspect of all of organizational functions. There are numerous vehicles for communicating safety matters including notices, posters, in-house journals, bulletins, information sheets, circulars, safety committee minutes, incident and near-miss reports, meetings and team briefings. However, these can only be used to best advantage if managers consider the overall process and its various components. These include communicators and recipients, content of safety messages and the opportunities for feedback. Opportunities for open and honest communication on safety should be encouraged so that employees are not encouraged to hide information on incidents. Equally, if employees are uncertain about any aspect of safety they should be confident to ask for advice.

Everything that occurs within an organization communicates its commitment or lack of commitment to safety. For example, poor housekeeping makes a statement about employee motivation and management control.

Managing the human resource

Organizations have increasingly recognized the need to actively manage safe patterns of behaviour and to inculcate positive attitudes to safety in their workforce (HSE, 1989a; Cox and Cox, 1996). A number of strategies and interventions have been developed to address these issues including attitude change programmes, behaviour modification and motivation packages. One of the aims of these interventions has been to enhance employee motivation towards behaving safely by attitude change. Aldridge (1976) has reviewed some of the available theories of motivation in an attempt to explain why accidents happen. Several other researchers have explored motivation in relation to safety but none have provided clear guidelines on how safety motivation can best be achieved. However, many organizations use reward systems as a way of reinforcing the idea that safety is as important as other areas of the business. Examples of such initiatives include allocations of awards (cars, pens, food hampers, etc.) or plaques and prizes for safety competitions. Some organizations use bonus or incentive schemes based on lost time, accident or incident performance. It has been argued (Petersen, 1971) that incentive schemes may cause under-reporting of accidents rather than actually reducing their number. Petersen (1971) has also argued that a critical approach to the assessment of incentives is necessary within the sponsoring organization and he raises the following issues for consideration.

1. Do employees have real control over the health and safety issues which affect them?
2. If an individual meets their health and safety objectives is there a genuine reward for them?
3. Can one ensure that rewards other than those for meeting health and safety objectives do not swamp the value of health and safety rewards?

If these issues are not taken into consideration and employees are neither operating to safe procedures, nor given the necessary tools and equipment to operate safely, then safety incentive schemes may go badly wrong. Some approaches to safety management and safety attitude change focus on employee behaviour, on the basis that inducing people to adopt certain behaviours might modify attitudes in the desired manner. A project in this area, carried out by one of the current authors (Cox, 1988b), used training workshops as a vehicle for attitude change. The key elements in this strategy are:

1.   The identification of relevant attitudes;
2.   Demonstration of the relevance of those attitudes as a guide to behaviour;
3.   Promotion of peer group pressure to facilitate change.

Behaviour modification techniques have been adapted to workplace safety by several organizations. These techniques which are used to shape or modify behaviour through the systematic application of reward have been derived from theories of conditioning and learning (Burkhardt, 1987). The approach taken by one of the current authors and her colleagues builds on these principles and utilizes a behavioural safety toolkit (Cox and Vassie, 1995).

Finally it is important to develop existing models of attitude change and to maximize their implications for safety. Traditionally, changing attitudes has been considered as an extension of the communication process in which a message is transmitted by one individual to another (or to a group) through a specified medium (see Figure 16.5). The aim of the communication is to influence the receiver. The elements of such a process are therefore:

1.   The individual or group of individuals who may change their attitude;
2.   The source of the argument or message;
3.   The content and context of the argument or message;
4.   The medium or channel of delivery.

Each of these elements is important in the attitude-change process. The source of the message must be credible and have status (for example, a senior manager or an 'expert'). The content of the message must have relevance and be meaningful. Face-to-face delivery is often most effective as is the use of the media (TV or newspapers). Individual characteristics (McKenna, 1987) of the receiver (including age, intelligence and personality) are also critical, and should be taken into account. Such models may be tailored to meet specific requirements with respect to safety.

Strategies for the reduction and management of errors

In previous sections we have discussed the application of risk management methods in the management of safety and have stressed the importance of actively managing the human resource. This final

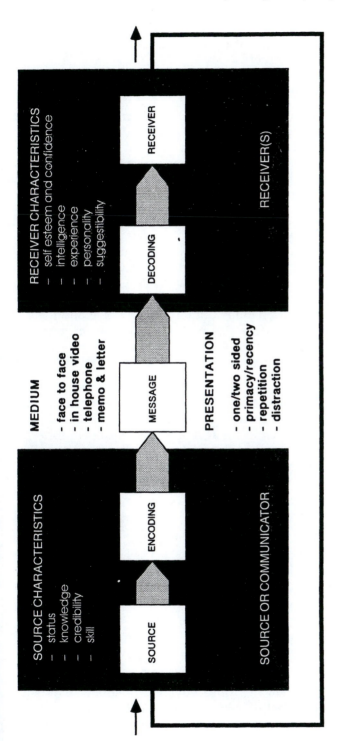

FIGURE 16.5
Communication model of attitude change

section briefly introduces four potential error handling strategies in the context of safety management. It draws on the human factors issues considered in previous chapters.

Intelligent decision support systems

Developments in computerized support systems may provide operators with additional information on the state of the system (Hollnagel *et al.*, 1988). For example, safety parameter display systems can show trends in important system state variables such as temperature and pressure. This may be particularly useful in complex systems and aid the decision making process outlined in Figure 16.2.

Memory aids for maintenance personnel

A disproportionate amount of accidents occur in maintenance activities (see Chapter 5). Reason (1990) has described the development of a portable interactive maintenance auxiliary (PIMA) which has been designed to form part of a maintenance technician's basic equipment in nuclear power plant installations. Although PIMA was designed as an external memory aid it may also be used as part of an ongoing monitoring system for health and safety.

Safety by design

A growing body of literature exists on design considerations (see Chapters 5 and 7). In the context of human error, designers and managers should assume that errors will occur and thus plan for error recovery. In particular they should make it easy to reverse operations and hard to carry out non reversible ones.

Error management

This is a procedure developed at the University of Munich by Michael Frese and his co-workers. It stemmed from empirical research on errors in human-computer interactions (Frese, 1987). They observed training errors and noted that these could have both positive and negative effects. The aim of error management is to promote the positive and to mitigate the negative effects of training errors in a systematic fashion (for further details see Frese, 1987; Reason, 1990).

Conclusion

This final chapter builds on the topics covered in earlier sections to present an integrated model of safety management. It recognizes the importance of good information sources and the need to incorporate safety and reliability considerations into system design.

It discusses the use of risk assessment techniques where appropriate and promotes a risk management approach. However, health and safety management is fundamentally good 'people' management. It requires organizations to consider the human resource and to develop systems which not only minimize accidents and incidents but also make better safety inevitable.

The consequences of poor or inadequate safety management can be disastrous. The final chapter of the book examines a number of 'high profile' disasters to illustrate some key systems management failures.

References

Aldridge, J. F. L. (1976) Safety and motivation, *J. Safety Surveyor*, September, 19.

Advisory Committee for Safety in Nuclear Installations (ACSNI) (1993) *Organising for Safety – Third Report of the Human Factors Study Group of ACSNI*, HSE Books.

Andreoni, D. (1986) *The Cost of Occupational Accidents and Diseases, Occupational Accidents and Diseases*, Occupational Safety and Health Series, No. 54, ILO, Geneva.

Bamber, L. (1983) Techniques of accident prevention, in *Safety at Work*, J. Ridley (ed.), Butterworth-Heinemann, London.

Bird, F. E. and Germain, G. L. (1987) *Practical Loss Control Leadership*, Institute Publishing, Loganville, GA.

Bryant, M. (1984) *Success with Occupational Safety Programmes*, Occupational Safety and Health Series, ILO, Geneva.

BS 5750 (1987) *Quality Systems: Specification for Design, Manufacture and Installation*, British Standards Institution, London.

BS7229 (1989) *Quality Systems Auditing*, British Standards Institution, London.

Burkhardt, F. (1987), 'A five-step method to modify behaviour in accident/incident concentrations', in *Successful Accident Prevention Programmes*, Menckel, E. (ed.), Arbete och Halsa, 32, 35.

CBI (1990) *Developing a Safety Culture*, Confederation of British Industry, London.

CHASE (1987) *The Complete Health and Safety Evaluation*, Health and Safety Technology and Management Ltd (HASTAM).

Checkland, P. B. (1981) *Systems Thinking, Systems Practice*, John Wiley, Chichester.

CIMAH (1984) *The Control of Industrial Major Accident Hazards Regulations 1984*, HMSO, London.

COSHH (1988) *The Control of Substances Hazardous to Health Regulations 1988*, HMSO, London.

Cox, S. (1988a) *Safety Training*, Health and Safety Officers Handbook, 6, 9.

Cox, S. (1988b) *Attitudes to Safety*, M.Phil. Thesis, University of Nottingham.

Cox, S. (1991a) Employee safety education and training in industry, in *The*

Proceedings of the First IUPAC Conference on Safety in the Chemical Industry, Basel, September 9–13.

Cox, S. (1991b), 'Bibliography of safety training materials', in *The Proceedings of the First IUPAC Conference on Safety in the Chemical Industry*, Basel, September 9-13.

Cox, S. and Cox, T. (1996) *Safety, Systems and People*, Butterworth-Heinemann, Oxford.

Crockford, G. N. (1980) *An Introduction to Risk Management*, Woodhead-Faulkner, Cambridge.

Crosby, P. B. (1979) *Quality is Free*, McGraw-Hill, New York.

Dawson, S. (1997) Corporate responsibility and effective management, in *Safety by Design, An Engineer's Responsibility for Safety*, (S. J. Cox, Lecture Coordinator) pp. 127–149, Hazards Forum.

Dawson, S., Willman, P., Clinton, A. and Bamford, M. (1988) *Safety at Work: The Limits of Self Regulation*, Cambridge University Press, Cambridge.

Fido, A. T. and Wood, D. O. (1989) *Safety Management Systems*, Blackmore Press, Shaftesbury.

Frese, M. (1987) A theory of control and complexity: Implications for software design and integration of computer systems into the workplace, in *Psychological Issues of Human Computer Interaction in the Workplace*, M. Frese, E. Ulich and W. Dzida (eds), Elsevier, Amsterdam.

Griffiths, D. K. (1985) Safety attitudes of management, *Ergonomics*, 28, 61.

Groeneweg, J. (1994) *Controlling the Controllable*, DSWO Press.

Hale, A. R. and Hale, M. D. (1972) *A Review of the Industrial Accident Literature*, Research Paper, National Institute of Industrial Psychology, London.

Hale, A. R. (1984) Is safety training worthwhile?, *J. Occup. Accidents*, 6, 17.

Hammer, W. (1981), *Handbook of System and Product Safety*, Prentice Hall, Englewood Cliffs, NJ.

Health and Safety Commission (1996) *Health and Safety (Consultation with Employees) Regulations 1996*, HMSO, London.

Health and Safety at Work etc. Act 1974 (1974) HMSO, London.

Hollnagel, E., Mancini, G. and Woods, D. (1988) *Cognitive Engineering in Complex Dynamic Worlds*, Academic Press, London.

HSE (1985) *Monitoring Safety HSE Occasional Paper Series*, OP9, HMSO, London.

HSE (1989a) *Human Factors in Industrial Safety*, HMSO, London.

HSE (1989b) *Quantified Risk Assessment: Its Input to Decision Making*, Health and Safety Executive, HMSO, London.

Juran, J. M. (1988) *Juran on Planning for Quality*, Free Press, New York.

Latimer *V* AEC Ltd (1953) AC643, 9153, 2, All ER, 449, HL.

Leplat, J. (1984) Occupational accident research and systems approach, *J. Occup. Accidents*, 6(2), 77.

Likert, R. (1967) *The Human Organization*, McGraw-Hill, New York.

Lowrance, W. W. (1990) *Stewardship of Chemical Production Risks*, Paper presented at the First IUPAC Workshop on safety in chemical production, September 9–13, Basel, Switzerland.

McKenna, E. F. (1987) *Psychology in Business: Theory and Applications*, Lawrence Erlbaum Associates, London.

Oliga, J. C. (1988) Methodological foundations of systems methodologies', *Systems Practice*, 1(1), 87.

Park, K. S. (1987) *Human Reliability, Advances in Human Factors/Ergonomics*, Elsevier, London.

Peters, T. J. and Waterman, R. H. (1982), *In Search of Excellence*, Harper and Row, New York.

Petersen, D. (1971) *Techniques of Safety Management*, McGraw-Hill, New York.

Raafat, H. M. N. (1989) Product liability and risk strategy, *J. Health and Safety*, 2, 19.

Reason, J. (1990) *Human Error*, Cambridge University Press, Cambridge.

Singleton, W. T. (1984) Application of human error analysis to occupational accident research, *J. Occup. Accidents*, 6(2), 107.

Steel, C. (1990) Risk Estimation, *The Safety and Health Practitioner*, June, 20.

Stranks, J. (1990) *A Manager's Guide to Health and Safety at Work*, Kogan Page, London.

The Regulations on Safety Representatives and Safety Committees 1977 (1977), HMSO, London.

Walters, D. and Gourlay, S. (1990) *Statutory employee involvement in health and safety at the workplace*: a report of the implementation and effectiveness of The Safety Representatives and Safety Committee Regulations 1977, HSE Contract Research Report No.20/1990.

Waring, A. E. (1989) *Systems Methods for Managers: A Practical Guide*, Blackwell, Oxford.

Whiston, J. and Eddershaw, B. (1989) Quality and safety – distant cousins or close relatives? *The Chemical Engineer*, June, 97.

Chapter 17

Some recent incidents and their implications

Introduction

Three Mile Island, Chernobyl, Bhopal, Mexico City, Zeebrugge and Kegworth are all place names which have become familiar to us through various media reports. Although they are many miles apart they share a common misfortune. They are places where major incidents and disasters have occurred; disasters in which many people lost their lives and the consequences of which are still being felt in terms of human pain and suffering. Many of the other costs of such disasters, social, environmental, economic and organizational, and their causes continue to be evaluated. Indeed causal analysis (see Chapter 4) is a well established part of accident investigation procedure and communication of causes of past incidents should, in theory, help prevent recurrence of similar incidents in the future. However, most major disasters provide us with ample evidence of the failures of organizations to learn from their own or other organizations' previous experiences. For example, in the case of the Three Mile Island incident a similar accident had occurred some months before at the technologically similar Davis Besse plant (Embrey, 1989). In the Davis Besse incident, correct operator action had prevented an accident (the operator had responded to the meta-stable system state, see Chapter 5). The Zeebrugge enquiry also revealed that there had been several occasions prior to Zeebrugge when ferries had left port with their bow doors open, but without incident in these cases. These examples provide a salutary reminder that organizations should respond not only to consequences but to antecedent conditions.

This chapter will review several major accidents and incidents and, using such case histories, will highlight a number of key points relating to:

1.   The importance of applying the methodology of reliability and risk assessment discussed in earlier chapters;
2.   The need to design 'forgiving' systems and the importance of an integrated systems approach;
3.   The importance of communication in the prevention of accidents and incidents;
4.   The human, economic, social and environmental costs of disaster.

It will also present a systems framework for accidents and incidents in line with previous chapters. The reader will be referred to additional references to support the case studies.

Systems models of accidents and incidents

A systems approach to the analysis of accidents and incidents recognizes the inter-relatedness of the various components of the accident process (see Figure 7.1) and accepts multicausality (Cox and Cox, 1996). It is rarely the case that accidents occur as a result of machine and equipment (hardware) failures alone. More often they occur as a result of a combination of organizational policy and procedures (software), human actions (liveware) and hardware failures. This approach is typified in the work of Embrey (1989) who has described a simple system-induced error model. This model is based on the hypothesis that all individuals have certain error tendencies. Furthermore, these error tendencies have to be combined with error-inducing conditions (organizational time pressures and deadlines, occupational stress, distractions, etc.) for an error to result. For the error to give rise to a significant consequence for either safety or reliability, an unforgiving environment has to exist. An unforgiving environment prevents or reduces the likelihood of error recovery described in Chapter 5. It also involves a metastable (vulnerable) system state, which in combination with the unrecovered error leads to the undesirable consequence. In some cases the error may give rise to a vulnerable state which does not produce a significant consequence, Embrey (1989) has termed this a latent failure.

It is interesting to note that many key actions with respect to safety have been consequence-driven and have ignored the latent failures or 'near-misses' which provide an invaluable source of information for safety and reliability. However, when such information is available it is often ignored or organizational decisions are made in light of other considerations rather than on technical reliability (see, for example, the Challenger case history below).

Case histories

Seven case histories are considered. They include the Challenger space shuttle, the Chernobyl RBMK reactor, Flixborough, Mexico City, Bhopal, Piper Alpha and the emergency landing of a Boeing 737. The sequence of events leading up to and including each 'disaster' is described. Each case history provides numerous learning points and key points are selected for further consideration.

The NASA shuttle 'Challenger'

On 28 January 1986 the shuttle Challenger exploded in a ball of fire 12 miles above the Atlantic Ocean off Cape Canaveral, Florida (Harris, 1986), killing all the crew. The tragedy was particularly poignant since a female school teacher was aboard the shuttle and thousands of American school children were watching for the first lesson from space. Three months after Challenger, a Titan rocket exploded over Vandenberg Airforce Base in California, and one month later, on 3 May 1986, a Delta rocket carrying a hurricane-spotting satellite had to be destroyed over Cape Canaveral. NASA, and the United States space programme in general, were in serious trouble as a result. How did such catastrophic system failures occur?

The Rogers Commission, the body charged with investigating the Challenger tragedy, made a number of observations (for an overview see Groves, 1986 and Cooper, 1987), many of which focused on the organizational problems, including management accountability, span of authority, complex reporting systems and the problems associated with meeting NASA's declared goals. The direct cause of the explosion, however, was a failure of one of the booster rocket's refractory lining O-ring seals. This rubbery seal split shortly after lift-off, releasing a jet of ignited fuel that caused the entire rocket complex to explode.

History of the O-ring
Investigations revealed the following sequence of events:

Jan 1979	The Chief of the Solid Motor Branch informed higher management that the O-ring seal was malfunctioning.
May 1980	NASA Engineering Panel noted that the O-ring seals of a shuttle booster rocket failed during a ground test.
Dec 1982	The Engineering Panel added the O-ring seal to its 'criticality' list and recorded the following statement '...it lacked a reliable back-up and, if the joint failed, it would lead to a loss of mission and crew'. (The Rogers Commission noted that in December 1982 there were a total of 748 parts on this list.)

Spring 1983	A formal launch constraint was placed on Challenger by the Booster Manager (this was routinely removed prior to each shuttle launch).
1984	(a) Further warning to superiors from Chief of Solid Motor Branch concerning O-rings. (b) Booster Manager receives letter from the Chief Engineer from United Space Boosters (a private company) highlighting similar problems. Eventually the top level management within NASA recognized they had a problem with the O-ring seal and asked Morton Thiokol (the prime contractors for the booster rocket) to seek a solution to the problem.
Jan 1985	A shuttle was launched following record low overnight temperatures and Thiokol engineers reported the most extensive O-ring damage ever.
Aug 1985	Thiokol engineers reported O-ring seal damage even at 50°F.
Oct 1985	The task force progress within Thiokol was hampered by internal pressures and the co-ordinating engineer approached his superiors for help.
Dec 1985	Thiokol's Special Projects Manager decided to take the O-ring problem off their priority list.
22 Jan 1986	Six days prior to the fatal 'Challenger' launch an unsigned memo was issued by NASA declaring 'this problem is considered closed'.
27 Jan 1986	A pre-launch teleconference amongst four Thiokol Vice-Presidents and the NASA senior managers to discuss the effect of the cold weather on O-ring reliability and the possibility of delaying the launch voted against such a delay (against the advice of engineers).

This chronicle of events has established beyond doubt that NASA was aware of the potential for failure associated with the O-ring seal. Why, therefore, did they decide to go ahead with the launch in the face of such evidence?

Jerome Lederer, founder of the private Flight Safety Foundation (Groves, 1986), offered this explanation: 'There was social pressure: they had thousands of school kids watching for the first school lesson from space. There was media pressure: they feared if they didn't launch, the press would report more delays. And there was commercial pressure: the Ariane (European launcher) was putting objects in space at much lower cost. NASA was also trying to show the Air Force that it could launch on schedule. The pressures were subtle, but they acted upon them'.

Official conclusions and recommendations

The basic recommendation made by the Rogers Commission (June, 1986) was that NASA should take firm control of its sprawling and decentralized bureaucracy and, moreover, cut back on meaningless paper flow. With respect to 'safety', the Commission recommended that anyone in the NASA system holding a strong view regarding a safety issue *must* be permitted to express their opinion *at any level*, rather than be limited to communicating their concerns only through departmental channels. Conversely, the Commission felt that top-level managers in NASA should take the initiative to raise oral questions, rather than simply send paper inquiries down through the ranks where they can be lost, ignored, or merely given lip service. In addition, the Commission recommended the adoption of new rules for launch. First, all pre-flight discussions of whether a launch should or should not go must be recorded. Second, the astronauts themselves should be involved in the decision-making process.

With reference to the redesign of the O-ring, the Commission recommended that the task should be supervised by independent experts. They also recommended that crew escape systems be investigated.

Learning points

The 'Challenger' case is an interesting illustration of the multifactorial causality of major accidents. It focuses attention not only on the technical problems associated with the O-ring seal but also on the communications in large and often 'sprawling' organizations like NASA. A formal quantified safety and reliability assessment of the booster rocket system, and particularly its performance in cold conditions, would provide an objective tool for decision making. However, the investment in redesign of this system could not fully solve the problem.

The case illustrates the complexity and irrationality of organizational decisions and provides an example of the cost-benefit process in which safety considerations are offset against other organizational goals.

Chernobyl

At 01.24 on Saturday, 26 April 1986, two explosions (about 3–4 seconds apart) blew off the 1000-tonne concrete cap sealing the Chernobyl-4 reactor, throwing radioactive material into the atmosphere. This was the worst accident in the history of commercial nuclear power generation and its impact on both lay and scientific communities world-wide has been immense (Mould, 1988).

The immediate cost of the accident was estimated at 30 lives, contamination of about 400 square miles of land around the Ukrainian plant (Besi *et al.*, 1987), and massive disruption of the community with the evacuation of 120 000 people. The longer-term effects of the accident are predicted to have increased the risk of cancer deaths in the immediate vicinity of the plant. However, the latent health effects in western Europe and Scandinavia may not be statistically significant when viewed against the normal mortality rate over the next 40 years. Additional socio-economic effects also include damage to the world agricultural industry and to the food chain (for example, United Kingdom sheep farmers were still being compensated for contaminated lamb in spring 1990).

The fatal accident sequence was initiated by a decision of the Chernobyl management and specialists to carry out an overnight experiment to test the ability of the turbine generator to power the emergency core cooling system (ECCS) whilst the generator was freewheeling to standstill after its steam supply had been shut off. This information would allow the management to determine whether the power requirement of Reactor 4 could be sustained for a short time during a power failure and until standby generators could be switched in. (It has been admitted: (1) that these tests were not properly planned, (2) that they had not received all the required approval and (3) that the safety instructions were minimal.)

The Chernobyl disaster occurred as the result of a number of factors including certain design features of the reactor, lack of planning and a series of deliberate violations of the operating rules by the plant operators (IAEA, 1986). How and why did a group of trained and competent operators commit the right blend of errors to blow up the reactor? In order to answer these questions we have to consider a number of key issues including the design of the reactor, the nature of the experimental work which the plant operators and engineers were involved in at the time of the accident and the sequence of events. (Much of the explanation for the accident concerns human factors and illustrates the themes discussed in Chapter 7.)

The RBMK reactor

The Chernobyl reactors are of the RBMK type. RBMK is a Russian acronym (*reaktor bolshoi moschnosti-kanalye*) for a high-power boiling water reactor. It was developed from a 1954 design (at Obninsk), and the concept is unique to the USSR, except that the Hanford-N reactor (in the United States of America) has similar reactor-physics principles. The first RBMK reactor of 1000 MW capacity came into operation in Leningrad in 1974. Four RBMK reactors were in operation on the Chernobyl site at the time of the accident and two more were under construction. The units were constructed in pairs, sharing common buildings and services. Unit 4 (coupled with Unit 3) became operational during 1984.

The RBMK reactor utilizes the energy released by fission of nuclei of uranium atoms to turn water into steam which, in turn, drives turbine generators to produce electrical power. It is cooled by circulating water that boils in the upper parts of vertical pressure tubes to produce steam. The steam is produced in two cooling loops, each with 840 fuel channels, two steam separators, four coolant pumps and associated equipment. The steam separators supply dry steam to two 500 MW turbo generators. A major part of the coolant circuitry is enclosed in a series of strong containment structures (IAEA, 1986).

Two design features of the plant are particularly noteworthy in the context of the accident:

1. The plant was designed at a time when the computing and control facilities (see Chapter 8) were relatively primitive and therefore much of the emergency response was manually driven;
2. The RBMK was known to be inherently unstable at low power and it was therefore forbidden to operate the reactor below 20% of maximum power. This involves the concepts of positive void coefficient and positive power coefficient which are described in the report on the Hinckley Point Inquiry (Barnes, 1990). It is sufficient for us to note this operating rule and to realize that compliance was dependent on the operator. There were no physical safeguards in the reactor to ensure compliance. At the time of the accident, No. 4 unit was operating at less than 20% of full power.

The sequence of events

A relatively short period of time had been scheduled for the experiment, immediately prior to an annual maintenance shutdown. The principal events occurred between midday on Friday 25 April and just after dawn on Saturday 26 April. They are presented below in chronological order and significant operator actions are highlighted (IAEA, 1986).

25 April 1986

01.00	Power reduction was commenced with the intention of achieving 25% power for test condition.
13.00	Reactor power reduced to 1600 MW. No turbine disconnected.
14.00	ECCS disconnected from primary circuit.
14.05	Kiev controller asked Unit 4 to continue supplying grid. The ECCS was not reconnected (*Major Fault No. 1*). (This represented a violation of written operating rules for just over 9 hours and may be a reflection of the operator's attitudes towards the safety of the plant.)
23.10	Unit 4 was released from the grid and continues power reduction to stabilize the reactor at between 700 and 1000 MW, or about 25% full power.

26 April 1986

00.28	On going to lower power an alternative set of control rods, called the automatic control rods (ACs), were switched in. However, the operators failed to reset the set point of the ACs (*Major Fault No. 2*) and were unable to prevent the reactors' thermal power dipping to 30 MW.
01.00	After a long struggle, the reactor power was finally stabilized at 200 MW, well below the intended level and well into the low power danger zone. (At this point the experiment should have been abandoned, but it was not. *Major Decision Error No. 1.*)
01.03–01.07	Two of the eight standby main circulation pumps, which had not been in operation, were started up. (The safety regulations normally prohibited such an operating mode.) The flow rate of water to the core was thereby increased and some pumps were operating beyond their permitted limits (*Major Fault No. 3*). The effect was to cause a reduction in steam formation and a fall in pressure in the steam drums.
01.19	The feedwater flow was increased threefold. This caused more control rods to be removed. The reactor should have tripped because of the low level but the operators had overridden the trip signals. This removed one of the automatic safety systems (*Major Fault No. 4*). The water in the cooling circuit was nearly at boiling point.
01.22	The shift supervisor requested a printout that indicated only six to eight rods remaining in the core. (It was strictly forbidden to operate with fewer than 12 rods.) The supervisor decided to continue with the tests (*Major Decision Error No. 2*). This was a fatal decision: the reactor thereafter was without brakes.
01.23	The steam line valves to No. 8 turbine generator were closed (*Major Fault No. 5*). This was to establish the conditions necessary for repeated testing, but it also disconnected the automatic safety trips. The steam pressure began to rise and flow through the core started to drop. Power increase ensued.
01.23.40	An attempt was made to shut-down the reactor by driving in the emergency shut-off rods, but they jammed within the warped tubes.
01.24	Two explosions occurred in quick succession. The sequence of events at this point is based on a

combination of visual observation, radiation measurement and calculations. The following facts are known.

01.30   Duty firemen were called out. Other units were summoned from Pripyat and Chernobyl.

05.00   Exterior fires had been extinguished. But the graphite fire in the core continued for several days. Only at this point was the neighbouring Unit 3 shut down.

Accident investigation and conclusions

The initial reports of the Chernobyl accident typified the media response to such events (see Chapter 12). For example, some of the British newspapers reported an immediate death toll of over 2000. The Russian delegate to the IAEA enquiry in August 1986 gave the following analogy of the accident (IAEA, 1986).

'Imagine personnel of a plane which is flying very high. Whilst flying they begin testing the plane, opening the doors of the plane, shutting off various systems...'. The facts (i.e. of the Chernobyl accident) show that even such a situation should have been foreseen by the designers (Legasov, 1986). He also made reference to the poor preparation for the experiment; safety measures were drafted in a purely formal rather than operational way and no provision was made for additional safety measures. The IAEA summary report on the post-accident review meeting highlights the following main contributory factors:

1.   Disabling of automatic trips (*Major Faults Nos 2 and 4*). Had these trips not been disabled, the insertion of the emergency rods would have terminated the transient regardless of all other circumstances;

2.   Operation at unacceptably low power levels (*Decision Error No. 1*). Following the initial power reduction, the reactor fell significantly below the minimum permitted level for continuous operation (700 MW). The experiment should have been terminated under these conditions;

3.   Additional actions (e.g. connecting additional pumps and increasing feed flow well above normal levels), created conditions for an accelerating power rise (*Major Fault No. 3*);

4.   Prior to tripping the turbogenerator, the feedwater flow was sharply reduced (*Major Fault No. 5*). The automatic rod system compensated for this reduction but did not have enough residual capacity to compensate for the reduction in main flow when the test started; without emergency shutdown protection the accelerating power rise was uncontrolled.

A systems approach to the event, however, recognizes the multicausality – the faults in the design concept of the reactor (hardware) and the failures in understanding and implementing the man machine interface concept. It also acknowledges the vital role of plant operating procedures (software), particularly when unusual operations are intended. The various actions of the personnel (i.e. management, 'experimenters' and operators) undoubtedly contributed to the accident; management should have controlled the experiment and provided the necessary technical support and operators should have followed defined procedures.

Reason (1987) has studied the psychological factors in the incident and highlighted several areas of research. The first concerns the cognitive difficulties people have in dealing with complex systems (see Chapter 7). The second includes the study of behaviour in groups.

We should also consider the problems of decision making under stress and its impact on the 'group'.

1. *The problems of coping with complexity* include insufficient consideration of processing time, difficulties in dealing with rapidly escalating developments, and thinking in causal series rather than causal nets. Doerner (1987) has used computer simulations to map out the strength and weaknesses of human minds in complex problem solving conditions and has noted all three difficulties. All three have relevance to the Chernobyl operators, but most especially the latter. When dealing with complex systems, people have a marked tendency to think in linear sequences. They are sensitive to the main effects of their actions upon the path to an immediate goal, but remain unaware of their side-effects upon the remainder of the system. In a tightly coupled, complex system, the consequences of actions radiate outwards like ripples in a pool; but people can only 'see' their influence within the narrow sector of their current concern.

2 and 3. *Group behaviour and decision making under stress.* How people make decisions under stress partly depends on whether they are working on their own or as part of a group. Groups may function to support decision making under stress by ensuring the flow of relevant information and by offering emotional as well as informational support to the individual. However, at the same time, group dynamics are consolidated and defended against contrary information or attempts to change them. 'Groupthink' may be a particular risk in very cohesive groups, or where there is a particularly strong and influential leader. Under stress, groups' members may expect stronger (and more authoritarian) leadership (Reason, 1984). The 'groupthink' syndrome has several other dimensions. In the Chernobyl incident it was exemplified in the 'illusion of invulnerability', 'the rationalizing away of anomalies', the unswerving belief in rightness of operations and the 'self-censoring' of doubts in their actions.

The individual under stress also works against a number of pressures in attempting to make correct decisions. For example, their attention appears to narrow in on to what are perceived to be the central (critical) aspects of the task. At the same time, their perception of the passage of time changes, and their levels of arousal increase possibly to supra-optimal levels. Under such conditions, it is likely that decision making will be impaired. This may possibly be because of the selection of the wrong information for decision making or of inappropriate responses, as well as impaired logic or the substitution of intuition or emotion for logic.

We should also acknowledge that, for the most part, humans do not plan and execute their actions in isolation, but within a complex social milieu. While errors may be considered in relation to the cognitive processes of the individual, the Chernobyl rule 'violations' can only be described with regard to the social context in which their behaviour was regulated (i.e. by operating procedures, codes of practice, rules and laws and the prevailing safety culture).

Learning points

The Chernobyl incident provides another illustration of the multicausality of accidents. The most important learning points are:

1. That intrinsically safe features should be incorporated into the basic design wherever possible. Lord Marshall (1987) has highlighted the limitations of the RBMK design;
2. That built-in safety features and interlocks should be provided where this is not possible. Reliance on instruction and training of the operators is just not adequate in such situations. Barnes (1990) has discussed the role of the Public Enquiry in ensuring adequate standards;
3. All accounts of the Chernobyl incident highlight the importance of strategies to minimize both the possibility and the consequence of human error. Training, management control, performance monitoring and quality checks on safety standards are all important;
4. Nuclear and other plants must be designed to accommodate man–machine interactions and should not place undue information processing demands on operators, particularly at times of stress.

More detailed analysis may be found in a report prepared for the Central Electricity Generating Board by Collier and Davies (1986).

Flixborough

The Flixborough works of Nypro (UK) Ltd was situated on a relatively isolated site surrounded by open fields. Two small villages are distanced about half-a-mile from the site and the nearest town is Scunthorpe which is three miles away. Between 1964 and 1967 a chemical plant was developed at Flixborough to make Caprolactam. This is a basic raw material used in the production of Nylon 6. Additional plant was installed between 1967 and 1972. This plant produced cyclohexanone by the oxidation of cyclohexane. The cyclohexanone was, in turn, used to manufacture Caprolactam.

In the new plant, the cyclohexane at a pressure of 8.6 bar and a temperature of 155°C, passed through a series of six reactor vessels by gravity feed. The oxidation took place as a result of the injection of air into the vessels. The output from the vessels was still about 94% cyclohexane which was subsequently separated out and recycled. As a result, a relatively large inventory of cyclohexane was present (about 120 tonnes). The vessels, made of half-inch mild steel with one-eighth inch stainless steel bonded liners were interconnected at 28 inch diameter apertures (Figure 17.1).

The incident
For some time before the explosion took place at Flixborough there had been no works engineer on site. Arrangements were under way to recruit a new works engineer following the departure of the previous one. In the interim period the site services engineer was acting in a co-ordinating capacity and advice and assistance were available from off-site. The report of the official enquiry into the Flixborough incident (Parker, 1975) was critical of the non-availability of adequate mechanical engineering expertise to deal with any complex or novel situations that might arise. The chronology of the incident was as follows:

27 Mar 1974   A cyclohexane leak was discovered from No. 5 reactor (see Figure 17.1). The plant was shut down for investigation.

28 Mar   Inspection indicated a 6-foot crack in the mild steel vessel. The leak indicated that the stainless steel

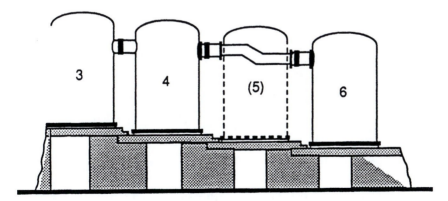

FIGURE 17.1
The cyclohexane oxidation vessels showing the position of reactor vessel 5 and the by-pass connection made after its removal

inner liner must also be damaged. Such damage to a reactor vessel normally operating at elevated temperature and pressure was clearly extremely serious. The nature of the damage was not understood. Despite this lack of understanding it was decided to keep the plant in operation by removing reactor number five and replacing it with a by-pass. This was necessarily dog-legged (Figure 17.1) and had a bellows unit at each end. The largest pipe available was of 20-inch diameter. This was used although the vessel apertures were of 28-inch diameter. No drawings were produced, calculations were inade-quate and the assembly was only tested to 8.8 bar, significantly below the 10.8 bar setting of the safety relief valves. The support system was insufficient to protect the pipework and bellows units against the shear forces produced by pressurization. Having completed the modifications, the plant was restarted.

29 May 1974   The plant was shut down to repair a minor leak.

1 June 1974   The plant was started up again. During this process the pressure reached 8.9–9.1 bar at one stage. (There was a reluctance to release the nitrogen pressurizing gas as this was in temporary short supply at the time.) Some hours later the 20-inch by-pass pipe ruptured either directly or as a result of a fire in an adjacent 8-inch pipe. The enquiry report (Parker, 1975) estimated that about 30 tonnes of cyclohexane escaped following the rupture and there resulted a massive unconfined vapour cloud explosion. On-site, 28 operators were killed and 36 were injured. Damage was extensive. The office block and control room were destroyed. Off-site, 53 were reported injured. Casualties would have been far worse on-site had it not been a Saturday when relatively few employees were present. The relative isolation of the site restricted injuries off-site.

Learning points

Important legislation followed the Flixborough disaster. The Advisory Committee on Major Hazards was set up in the United Kingdom in late 1974. This Committee produced three very important reports (Harvey, 1976, 1979, 1984). The work of the Committee led directly in the United Kingdom to the Notification of Installations Handling Hazardous Substances Regulations (1982) and also had considerable influence on the contents of the 1982 European Commission Major Hazards Directive

(EC, 1982). This in its turn was implemented in the United Kingdom in the form of the Control of Industrial Major Accident Hazards Regulations (1984).

Lees (1980) has listed many learning points from the Flixborough incident. These include:

1. The requirement for a high standard of management and technical expertise, which must be available at all times. Poor standards were evidenced by the unprofessional approach adopted to the problems of reactor vessel 5;
2. The necessity for adequate design and testing and the correct use of standards and codes of practice. British Standard BS3351 (BS, 1971) and the bellows manufacturers guidance notes would both have provided information as to how the bypass should have been correctly configured;
3. The importance of inventory limitation where hazardous chemicals are in use;
4. The need to take steps to limit exposure of personnel to potential hazards. A blast-proof control room could have saved many lives;
5. The particular care needed when plant modifications are made;
6. The importance of understanding and managing the potential conflict of priorities between safety and production.

Operations should not have re-started without a full investigation of the failure in reactor No. 5. Again re-start should have awaited delivery of fresh supplies of nitrogen so that pressure could be controlled without concern about nitrogen supplies being exhausted. The Flixborough incident will be remembered as a serious tragedy which could in other circumstances have been very much more serious. The Court of Enquiry was not without a degree of controversy over some technical matters, but it provided a very thorough and open investigation into the incident.

Mexico City

San Juanico, where this tragedy took place, is a settlement of 35 000 people outside Mexico City. The fire and explosions occurred on a site run by the state-owned PEMEX organization. The site was used for the storage and distribution of a liquefied petroleum gas known as LP-Gas, which was 80% butane and 20% propane. The LP-Gas was delivered to the site from distant refineries by one 12-inch and two 4-inch underground pipelines. Two companies which distributed the LP-Gas had depots immediately next to the PEMEX site.

The plant on the PEMEX site was originally built to American Petroleum Institute Standards, starting in 1961. Subsequent shortage of space had led to excessive crowding of the 48 cylindrical and six large spherical storage tanks. At the same time, lack of adequate planning control had resulted in habitations being constructed only 130 metres from the site. An account of the incident prepared by the Skandia International Insurance Company (Skandia, 1985) reports that the health and safety committee at the PEMEX plant had strongly criticized maintenance standards on several occasions. Pearce (1985) reports that there had been several incidents at PEMEX sites in the recent past in which 89 people had died and hundreds had been wounded. Pearce also quotes the findings of a team from the Netherlands Organization for Applied Scientific Research (TNO) that one of the booster pumps on the pipeline was capable of delivery at a pressure well above the design pressure of the storage tanks. There is thus clear evidence of poor safety standards, a history of previous incidents and inadequate design.

The incident
We follow the chronology given in the Skandia report (Skandia, 1985) on 18–19 November 1984.

18 November 1984
During the afternoon the filling began of the storage cylinders and spheres which at that time were almost empty. The liquefied LP-Gas was delivered by pipeline from a refinery 400 km away. By late evening, the two largest spherical vessels had been filled to about 90% of their capacity of 2400 m³. During the night the cylindrical tanks were similarly filled. The remaining four spheres, each of 1500 m³ capacity, had been half-filled when the incident occurred. It is estimated that 11 000 m³ of LP-Gas was on-site at the PEMEX depot at that time.

19 November 1984
05.35   A rupture occurred in the pressurized system in the vicinity of the storage vessels. The resulting gas cloud drifted slowly and spread to cover an area approximately 150 m by 200 m.

05.40   The gas cloud ignited causing both on-site and off-site damage.

05.45   The first boiling liquid expanding vapour (BLEVE) explosion took place as a result of the flames playing on a storage vessel.

05.46   One of the most violent explosions resulted from a BLEVE in one or two of the smaller spherical vessels. There was a 300 m diameter fireball and droplets of LP-gas fell on the adjacent housing areas, vaporized

	and caught fire, causing deaths, injuries and extensive fire damage.
06.00	The first fire fighters arrived.
06.30	There was traffic chaos as panicking evacuees obstructed the movement of emergency vehicles.
07.00	Direct television broadcasts from the site of the emergency are likely to have contributed to panic.
07.01	Explosions continued. The last of nine BLEVEs which were large enough to register on a seismograph at the University of Mexico, was recorded.
07.30	Tank explosions were still occurring. The fire department was beginning to get some of the fires under control however.
11.00	The last major tank explosion occurred.
23.00	Neither of the two 2400 m^3 spheres exploded. Such an explosion would undoubtedly have led to even greater death, injury and destruction. Both burned out without an explosion, the last flames going out at about 23.00 hours.

Casualties and damage

At least 500 people were killed in the incident and more than 7000 were injured, according to Skandia (1985). Damage was extensive due to both fire and missiles. One 20-tonne cylindrical vessel was projected 1.2 km and caused heavy damage to a two storey house where it landed. A total of 39 000 people were evacuated or made homeless.

Learning points

The Skandia publication (Skandia, 1985) quotes the Dutch (TNO) report on the lessons to be learned. The main points were:

1. The importance of site layout and of providing adequate spacing of chemical plant. The rapidity with which successive explosions took place was considered to be due to the close spacing of the storage vessels;

2. The importance of maintaining an open area around high-risk sites. There were houses only 130 m from the storage vessels at San Juanico. A spacing of 400 m would have avoided danger from the fireballs and from droplets of liquefied LP-Gas. Even this spacing would not be out of missile range;

3. The importance of maintaining high engineering standards and adequate maintenance levels;

4. The requirement for LPG plants to install adequate instrumentation including gas alarms, so that faults can be diagnosed rapidly and leakages minimized.

Bhopal

The factory at Bhopal in India was used to produce the pesticide carbaryl. The factory was set up in 1969 on a site outside Bhopal. At the time the location was relatively isolated, and in any case was used merely to formulate, package and ship the pesticide from materials bought in. At a later stage, methyl isocyante (MIC) was brought in, converted into the pesticide, formulated, packaged and shipped. In 1981 production of MIC on site was introduced. By this time the Bhopal plant of Union Carbide India Limited had been converted into a full-scale high-risk installation. Meanwhile a shanty town had grown up close to the site.

MIC production and storage

Union Carbide has claimed that the MIC production plant was designed to the same safety standards as similar plant in the United States (Browning, 1985). Indian operating staff received hands-on training in the United States before the Bhopal plant was commissioned and the commissioning was assisted by experts from the United States. (However, with the scale of the Bhopal disaster it is difficult to find unbiased sources.)

The highly toxic MIC is produced by reacting methylamine with phosgene. At Bhopal, the MIC, dissolved in chloroform, was stored in relatively large quantities in three storage tanks. It was then used to produce carbaryl. MIC is a highly reactive chemical and must be stored under closely controlled conditions. The Bhopal storage vessels were provided with a refrigeration system. The vessels were protected by relief valves and bursting discs, the output passing through a vent gas scrubber using caustic soda to neutralize the MIC. Finally, a flare stack was provided to burn off flammable gases or vapours. The scrubber system and the 33 m high flare stack were designed to deal with minor leaks. They were not able to handle large scale emissions.

Standard operating procedures required that all these safety features should be available when the MIC was in storage. Browning (1985) reports that an audit in 1982 indicated that the general state of the plant was satisfactory and recommended only relatively minor changes.

The incident

The sequence of events on the night of 2–3 December 1984 has been described by Kharbanda and Stallworthy (1988):

22.15   The shift supervisor asked an operator to wash the pipework in the vicinity of one of the three MIC storage tanks. The tank valves had been known to leak, so a slip blind was inserted to seal the tank. This

	was important as water entering the tank would initiate a chemical reaction that would produce heat.
23.00	The new shift came on duty and noted a pressure rise in the same vessel. This was assumed to be due to nitrogen pressurization used to transfer MIC out of the vessel.
23.30	The operators experienced irritation to the eyes due to an MIC leak. This happened every so often and no action was thought necessary.
24.00	Both temperature and pressure were found to be rising. Water was sprayed onto the vessel to no avail. Eventually the pressure relief system operated and about 30 tonnes of MIC were given off as a gas or vapour.

The MIC was transported by the wind in the direction of Bhopal. Most of the town was affected by a toxic cloud that covered an area approximately 5 km downwind by 2 km across. The precise number killed by the MIC is not known but it reached 2000 and possibly 3000. The number sustaining permanent injuries was 200 000 or more. At the time, little was known about the effects of large doses of MIC on humans and there was some controversy over treatment.

The incident was investigated by a team from the Union Carbide Corporation, United States of America (Union Carbide, 1985). The team took samples from the residue in the affected vessel and a number of experiments were performed in an attempt to reproduce similar conditions. They concluded that:

1. Between 120 and 240 gallons of water had entered the vessel. The report ruled out the washing operations as the source of the water and suggested that the introduction had been a deliberate act;
2. The temperature had reached 200°C or more and that reactions involving 40% of the MIC would have led to the vaporization of the remainder of the MIC;
3. There had been up to 5% chloroform present, significantly more than the percentage dictated by standard operating procedures. This had produced additional catalytic effects at the high temperatures that were generated;
4. Neither the refrigeration system, the scrubber nor the flare were in operation at the time of the incident, contrary to standard operating procedures.

A number of other points can be made. For example, there was no need to hold such a large inventory of MIC at Bhopal. It would have been very much safer to reduce the inventory and to convert the MIC more rapidly into the relatively harmless carbaryl. Again the local inhabitants had been given no information about the hazardous nature of the operations involved and no emergency procedures had been set out and explained. This was particularly serious in view of the close proximity of housing developments to the Union Carbide site.

Learning points
Bhopal provides another example of latent failures (Embrey, 1989). Poor management systems, botched maintenance, operator errors and bad governmental decisions are all evident in this case.

The main learning points from the tragedy are:

1. The importance of strict adherence to standard operating procedures;
2. The need for careful control of development both on- and off-site in order to ensure adequate separation between hazardous plant and local habitation;
3. The requirement for the local population to be provided with adequate information and an emergency plan;
4. The special duty of care on multinationals operating in countries where safety standards and controls are weak. The town of Bhopal had only two safety inspectors neither of whom had any qualifications or experience in chemical engineering;
5. The need for well defined safety management systems.

Piper Alpha

Piper Alpha was an oil platform located in the North Sea about 110 miles north-east of Aberdeen, Scotland. Its function was to drill for oil and gas which it extracted from the Piper oil field. The platform had facilities to separate and remove water from the crude oil, then to extract the hydrocarbon gas. This in turn was separated into non-condensable gas (mainly methane) and condensable gas (mainly propane). The non-condensable gas was normally piped to the St. Fergus gas terminal on mainland Scotland but could be flared-off when necessary. The condensable gas was compressed and liquefied and then added to the oil stream which was piped directly to Flotta in the Orkney Islands, off the north coast of Scotland.

Piper Alpha was connected to other neighbouring platforms. It received gas from the Tartan platform. This was sent, together with Piper Alpha's own output, to the MCP-Q1 platform and onward to St. Fergus. Gas was also sent from Piper Alpha to the Claymore platform.

The Piper Alpha catastrophe in which 167 persons died took place late in the evening of 6 July 1988. The platform was particularly busy that day and three conventional vessels and a semi-submersible vessel were in attendance. Normal production was under way but, in addition, extensive modification work was taking place, as was routine maintenance on various items of plant and equipment. In particular, condensate injection pump A, one of the two pumps used to return the liquefied condensate to the oil pipeline for transmission to Flotta, was out of operation for preventative maintenance under a formal Permit to Work system. Pump B was in normal use.

The public enquiry into the Piper Alpha incident was chaired by Lord Cullen (Cullen, 1990). The enquiry faced a difficult task in determining the course of events. Many potential witnesses lost their lives in the tragedy while most of the physical evidence was missing because of the collapse and loss of the rig. The most likely course of events on the evening of 6 July and the early hours of 7 July 1988 was:

21.45	Condensate injection pump B tripped out. Unsuccessful attempts were made to re-start it. Maintenance work on pump A had been suspended overnight and the decision was taken to re-start and use it. The Permit to Work was duly signed off and action was initiated to de-isolate pump A and bring it into operation. The shift operators were unaware that the only pressure safety valve (PSV 504) had been removed from pump A during the day and replaced with a blanking flange which had not been tightened up. This vital information was not mentioned either on the Permit document or verbally on shift handover. Starting the pump led to the release of some tens of kilograms of condensate.
22.00	The initial explosion caused by the ignition of the condensate by some unknown heat source was followed within seconds by the production of a fireball fuelled by a massive leak of crude oil resulting from the rupture of a pipe. The initial explosion put the main power supplies out of action and the emergency systems were largely ineffective. Gas detector alarms sounded only seconds before the explosion. The fire water system failed to operate. Lighting in the accommodation modules, where many of the operatives were located, was lost. Emergency lighting came on but failed after 10–15 minutes. The emergency shutdown system operated but there was evidence that the emergency shutdown valve on the main oil line to Flotta failed to close fully. This further fuelled the crude oil fire. Over the next 20 minutes 22 survivors left the platform, mainly from the lower work-levels.
22.20	There was a major explosion due to the rupture of the riser on the pipeline bringing gas to Piper Alpha from Tartan. It was followed by a high pressure gas fire generating intense heat.

22.50	About 39 survivors had now left the platform. There was a further massive explosion caused by the rupture of the MCP-01 gas riser. Structural collapse due to the high temperatures involved now started. By 00.45 the platform had almost completely disintegrated.
23.20	Following a number of further explosions, the Claymore gas riser ruptured, further hastening structural collapse.
23.30	By this time, 62 survivors had escaped from Piper Alpha. There was complete confusion in the accommodation modules where many operatives had gathered. Evacuation procedures were unusable due to smoke and flames. No effective control was exercised and no instructions were given. Those that escaped did so by their own efforts, some jumping off the platform into the sea, others lowering themselves down ropes and hosepipes. Those rescued were transferred to the semi-submersible platform, Thoros.
02.02	The first casualties left Thoros by helicopter for hospitalization. One of the 62 subsequently died in hospital. All survivors had reached the shore by 08.15. The disaster claimed the lives of 167 persons.

Findings of the enquiry
The enquiry report made a number of adverse comments about the safety regime on Piper Alpha:

1. It drew attention to the failures of the Permit to Work System in operation on the platform, pointing out that the system had already been criticized on a previous occasion. In particular, there were no lock-off procedures to prevent unauthorized re-starting of equipment and there was a lack of a clear mechanism for passing on vital information from shift to shift. Personnel were not provided with adequate training in the use of Permits to Work and this was made worse by the lack of enforcement of agreed procedures.

2. It criticized the practice of keeping the diesel fire pumps on manual control at certain times, despite the fact that this procedure had already received adverse comment in a safety audit. Even if the pumps had been started there was evidence that the deluge system had not been adequately maintained and would not have functioned properly.

3. Poor general standards of training for emergencies. In particular, the training in the use of emergency safety and escape equipment was inadequate.

4. There was clear evidence of the lack of involvement of senior management in critical safety matters – insufficient deter-

mination to ensure that agreed procedures were used, lack of involvement when problems arose, lack of realistic planning for major emergencies and an absence of any systematic attempt to assess and control major hazards.

The report also pointed out inadequacies in the Department of Energy's inspection procedures. The inspectorate was clearly affected by persistent under-manning and the inspections made much use of sampling techniques. They had not picked up the many shortcomings of the Piper Alpha safety management system. This was obvious if the findings of inspections in June 1987 and June 1988 were compared with those of the enquiry.

Learning points

A number of important recommendations regarding offshore safety management were made in the enquiry report. Of particular importance was the proposal to introduce a requirement for formal safety assessment of the major hazards on each installation, the findings to be presented in the form of a written safety case. This document was to demonstrate how risks had been identified, the nature of the control measures to be employed and the adequacy of these measures in providing safe working conditions. Particular measures recommended were the provision of temporary safe refuges, together with adequate escape routes and embarkation points and the provision and maintenance of effective safety management systems.

The report recommended the transfer of enforcement duties from the Department of Energy to the Health and Safety Executive. The recommendations became the basis for the Offshore Installations (Safety Case) Regulations 1992. Piper Alpha also shows up the importance of:

1. Having a comprehensive safety management system in place, involving all staff from the most senior to the most junior, providing safe systems of work, providing adequate standards of training, instruction and supervision and ensuring that standards are maintained and improved by use of monitoring and review;
2. Using risk assessment as a method of generating adequate standards of self-regulation;
3. Employing sufficient training and supervision to ensure that Permit to Work systems do, in fact, provide safe systems of work in high-risk environments;
4. Providing realistic workable procedures for dealing with large-scale emergencies.

Emergency landing of a Boeing 737

Most of the incidents described in this chapter were well-publicized and involved multiple loss of life. The emergency landing in the present case was entirely successful – there were no deaths and no injuries. The incident was chosen for inclusion as it illustrates how closely reliability and safety can be linked in some circumstances, and how important it is to ensure the highest maintenance standards when this is so.

The incident involved a Boeing 737-400 based at East Midlands Airport in England. It was due to undergo routine 750-hour engine checks. These took place overnight under far from ideal conditions and the aircraft was not correctly reinstated ready for its next flight to Lanzarote Airport in the Canary Islands next morning. As a result there was a rapid loss of oil at early stages of the flight and catastrophe was only averted by the rapid and effective actions of the flight crew.

On the night in question the line maintenance team had two of its six members absent and, of the other four, two, including the shift leader, were working extra nights to cover for other absences. At the same time, the base maintenance controller was operating with four of his five supervisors either on leave or missing due to illness. The incident occurred overnight on 22 and 23 February 1995. The sequence was as follows:

19.30	The line engineer, due to undertake the maintenance work, arrived at work to find that there had been no response to his request for extra manpower. He started the work but later handed over to the base maintenance controller who offered to undertake the task because it would ensure continuation of his authorization to perform this particular operation. There was no written handover, purely a verbal one and, in any case, no suitable proforma was available for such a procedure. The controller, having taken over, did not make use of descriptive task cards which were readily available. He made reference to his own training notes but these were not comprehensive. The work was interrupted several times while the controller dealt with other matters. Almost inevitably, mistakes were made. Two rotor drive covers, one on each engine, were left off, and the engines were not given a ground run test. Despite these shortcomings the maintenance work was signed-off in the technical log as complete.
07.30	The live engineering day shift leader noted from the technical log that the work was complete and the aircraft was prepared for flight. On handover to the air crew it was noted that a hatch had been left open and that two sets of circuit breakers had not been reset. These matters were put right. Later in the morning, luggage was loaded, passengers boarded and the aircraft took-off for Lanzarote.
12.05	The aircraft was still climbing to cruising altitude

when there was an indication of loss of oil pressure. The flight was diverted to Luton Airport and the engines were shut-down during the landing roll. There were no casualties, thanks to the prompt action of the air crew and ground control.

A subsequent company enquiry revealed that it was not uncommon to ignore recommended procedures in performing this particular engine check. There was a quality assurance system in place but this had not detected these shortcomings; staffing limitations were such that the QA department could only act as an administrative centre for documentation – no audit or inspection services were provided. Likewise, the Civil Aviation Authority's monitoring system had failed to note the procedural lapses. The official Air Incident Report (HMSO, 1996) noted these short-comings. It also criticized the way maintenance operations were being undertaken with insufficient staff – a monitoring system should have been in place and adequate staff provided. Such a system would need to take into account that staff concentration and reasoning ability are likely to be limited on night shifts when much of the maintenance work is undertaken. The report noted that there had been eight previous instances where the rotor drive covers had been left off at other airports – subsequent procedural changes had clearly not been effective. The report also drew attention to similarities with occurrences to two other United Kingdom aircraft.

Learning points
The most important learning points are:

1. The highest standards of plant maintenance must be in place when serious consequences may follow from equipment failure;
2. Workload must be carefully monitored in such circumstances to ensure that adequate staffing levels are maintained;
3. An independent and effective quality assurance system is needed in order to ensure correct procedures are adopted and adequate standards maintained.

Conclusion

The case studies described in this chapter provide illustrations of systems failures in a variety of socio-technical systems. These failures were not limited solely to failures in the technology, rather they illustrate how systems failures occur as a result of a combination of factors including failures in machine and equipment, human 'error' and lack of adequate organizational systems.

The reliability principles and risk assessment methodology as discussed in the first part of the book could (and should) have been applied to each of the situations. In those cases where they were applied, follow-up actions should have been taken to ensure the continuing safety and reliability of the system. Such actions are all part of the practice of health and safety management (see Chapter 16). It is only by adherence to these principles that such accidents may be prevented.

Further reading

Kletz, T. A. (1990) *Learning from Accidents in Industry*, Butterworth-Heinemann, Oxford.

Mosey, D. (1990) *Reactor Accidents, Nuclear Safety and the Role of Institutional Failure*, Butterworth-Heinemann, Oxford.

References

Barnes, M. (1990) The Hinkley Point Public Inquiries, Volume 6, *The Risk of Accidents*.

Besi, A., Kalfsbeek, H., Mancini, G. and Pancet, A. (1987) *Preliminary Analysis of the Chernobyl Account*, Technical Report No. 1.87.03 PER 1249 Ispra Establishment, Commission of European Communities Joint Research Centre, Italy.

Browning, J. B. (1985) After Bhopal, in *The Chemical Industry after Bhopal, an International Symposium*, London, November 1985.

BS (1971) BS 3351, *Specification for piping systems for petroleum and petrochemical plants*, British Standards Institution, London.

Collier, J. G. and Davies, L. M. (1986) *Chernobyl: The Accident at Chernobyl Unit 4 in the Ukraine, April 1986*, Central Electricity Generating Board, Barnwood.

Cooper, H. S. F. (1987) Letter from the Space Center, *The New Yorker*, November, p. 10.

Cullen, Lord (1990) *The Public Enquiry into the Piper Alpha Disaster*, HMSO, London.

Doerner, D. (1987) On the difficulties people have in dealing with complexity, in *New Technology and Human Error*, J. Rasmussen, K. Duncan and J. Leplat (eds), John Wiley, Chichester.

EC (1982) Major Hazards (Seveso) Directive European Communities 82/501/EEC.

Embrey, D. (1989) *The management of risk arising from human error, Conference Proceedings, Human Reliability in Nuclear Power*, Confederation of British Industry, London.

Groves, G. V. (1986) Shuttle Events, *Spaceflight*, **28**, July–August, 292.

Harris, G. L. (1986) Challenger Flight Record, *Spaceflight*, **28**, March, 102.

Harvey, B. H. (1976), *First Report of the Advisory Committee on Major Hazards*, HMSO, London.

Harvey, B. H. (1979) *Second Report of the Advisory Committee on Major Hazards*, HMSO, London.

Harvey, B. H. (1984) *Third Report of the Advisory Committee on Major Hazards*, HMSO, London.

HMSO (1996) *Report on the Incident to Boeing 737-400, G-OBMM near Daventry on 23 February 1995*, Air Incident Report 3/96, HMSO.

IAEA (1986) *Summary report on the post-accident review meeting on the Chernobyl Accident,* Vienna.

IBC Technical Services, London, EC (1982), *Major Hazards (Seveso) Directive,* European Communities 82/501/EEC.

Kharbanda, O. P. and Stallworthy, E. A. (1988) *Safety in the Chemical Industry, Lessons from Major Disasters,* Heinemann.

Lees, F. P. (1980) *Loss Prevention in the Process Industries,* Butterworth-Heinemann, Oxford.

Legasov, V. (1986) *Report of the USSR Delegation Leader,* 25–29 August 1986, IAEA, Vienna.

Lord Marshall of Goring (1987) *Nuclear Technology International,* N. R. Geary (ed.), Sterling Publications, London.

Michaelis, A. R. (1985) High technology accidents, unpredictable and inevitable, *Interdisciplinary Science Reviews,* **5,** 79.

Mould, R. F. (1988) *Chernobyl: The Real Story,* Pergamon Press, Oxford.

Parker, R. J. (1975) *The Flixborough Disaster,* Report of the Court of Enquiry, HMSO, London.

Pearce, F. (1985) After Bhopal, who remembered Ixhuatepec?, *New Scientist,* 18 July, 22.

Reason, J. (1987) The Chernobyl errors, *Bulletin of the British Psychological Society,* GOP, 201.

Report of the Presidential Commission on the Space Shuttle Challenger Accident (1986), Government Printing Agency, Washington, DC.

Skandia (1985) *BLEVE. The Tragedy at San Juanico,* Skandia International Insurance Corporation, Stockholm.

Union Carbide (1985) *Bhopal methyl isocyanate incident investigation team report,* Union Carbide Corporation, Danbury, Conn., March.

Postscript

In recent years industry has had to survive under increasing economic pressure. Such pressure has been generated at both national and international levels and has been associated with large scale restructuring. For example, many of the larger organizations have reduced the size of their workforce by removal of layers of management (downsizing and delayering) and by contracting out (outsourcing). Teams of workers are frequently expected to take on an extra workload and provide additional services which have required a broader range of skills (multi-skilling). Work teams have also had to take greater responsibility for their activities (empowerment). Such changes can significantly impact on operating costs and at the same time allow the organization to be more flexible in adapting to changing market demands and customer needs.

It is highly relevant for our study and understanding of health and safety management to enquire whether any of the serious incidents that have taken place in recent years had industrial restructuring as a contributory cause. In fact, such enquiries are fraught with difficulties. Although accident investigation reports try to provide an accurate account of how certain accidents came about, they do not normally provide insights into organizational change. Why, for example, were training and supervision standards so poor on the Piper Alpha rig, or why were the maintenance crews so under-staffed and the quality control team so ineffective in the Boeing incident (see Chapter 17)?

A recent study by the HSE in the United Kingdom (Wright, 1997) has identified two incidents, both involving multiple loss of life, which were clearly associated with large-scale organizational change. The study is of particular significance, however, in demonstrating how health and safety standards can not only be maintained but can actually be enhanced under such circumstances. It is based on the examination of ten widely differing organizations which had undergone large scale restructuring.

Wright (1997) has listed the following key health and safety issues to be addressed:

1   Senior management have to show real commitment to an agreed policy for change;
2   The impact of proposed organizational changes needs to be assessed at an early stage;
3   The competence of staff and contractors should be examined And all training requirements should be determined;
4   Responsibilities and accountabilities for health and safety should be clearly specified;
5   The level of health and safety resources required in-house needs to be matched to the risks involved;
6   The status of key safety rules and procedures needs to be settled;
7   The management of outsourcing should be given particular attention;
8   Care should be taken to ensure that adequate capability to cope with emergency situations is retained;
9   Full account should be taken of the impact of change on stress and morale;
10 Changes should be phased and managed with care and good standards of communication maintained with the workforce;
11 The impact of changes to be measured and monitored, and progress reviewed.

Changes involving devolution of responsibility to lower levels of management and the broadening of individual responsibility may be accompanied by heightened employee stress and anxiety. The situation must be managed with care if an effective and relatively stress-free transition is to take place. The devolved responsibilities will usually include health and safety management. Indeed, the management of health and safety can be expected to become less centralized and more evenly diffused through the organization although expert backup must be readily available. Training for new roles will be essential.

The actions listed are clearly in line with the health and safety management principles set out in earlier chapters. The issue here is to ensure that when significant changes are made to working methods, that corresponding changes are made to how health and safety is managed. No organization is static and review at one level or another can be expected to be an almost continuous process.

References

Wright, M. S. (1997) *Business Re-Engineering and Health and Safety Management*, HSE Books, London.

Appendix

(1)   Failure density function, hazard rate and reliability
We assume that there are n equipments and that at time t, $n_s(t)$ have survived and $n_f(t)$ have failed, such that $n = n_s(t) + n_f(t)$. If at time $t = 0$, $n_s(0) = n$ and $n_f(0) = 0$, then the definition of reliability, $R(t)$, is

$$R(t) = n_s(t)/n = 1 - n_f(t)/n \tag{A1}$$

Similarly, the unreliability $F(t)$ is

$$F(t) = n_f(t)/n \tag{A2}$$

The failure density function $f(t)$ is

$$f(t) = \frac{1}{n}\, dn_f(t)/dt \tag{A3}$$

while the hazard rate $z(t)$ is

$$z(t) = \frac{1}{n_s}\, dn_f(t)/dt \tag{A4}$$

From these expressions we find

$$z(t) = f(t)/R(t) \tag{A5}$$

$$R(t) = \div \int_t^+ f(t)dt \tag{A6}$$

$$F(t) = \int_0^t f(t)dt \tag{A7}$$

The mean time to failure, or mean lifetime, is

$$\int_- t\, f(t)dt \tag{A8}$$

(2)   The normal distribution
The failure density function for this distribution is

$$f(t) = \frac{1}{\sigma(2\pi)} \exp[-(t-m)^2/2\sigma^2]$$

where
    m = mean, i.e. the value of t at the peak, and
    σ = standard deviation, measuring the width of the peak.

Substitution of $t = m \pm \sigma$ and $t = m \pm 2\sigma$ shows that $f(t)$ falls to 60.6% and 13.5% of the peak value at these points. $R(t)$ and $z(t)$ are best evaluated from equations (A6) and (A7), respectively, either by numerical integration or from published tables.

(3)   The exponential distribution
In this case we have

$$f(t) = \lambda \exp(-\lambda t)$$

$$z(t) = \lambda$$

$$R(t) = \exp(-\lambda t)$$

Thus we confirm that $z(t)$ is a constant, independent of t. Note also that at $t = 0$, $f(0) = z(0) = \lambda$, as illustrated by the example in Chapter 2. As expected, $R(0) = 1$, while $R(t)$ and $f(t)$ both have the same exponential form.
The mean lifetime according to equation (A8) is

$$\int_{-\infty} t\, e^{-\lambda t}\, dt = \frac{1}{\lambda}$$

Further simplification is found where $\lambda t \ll 1$, that is at times very small compared to the mean lifetime, when

$$R(t) = 1 - \lambda t \text{ and}$$
$$F(t) = \lambda t \qquad\qquad\qquad\qquad\qquad\qquad (A9)$$

Unreliability

$$= (1/\text{mean lifetime}) \times \text{time}$$
$$= (\text{mean failure rate}) \times \text{time}$$

(4)   Fractional dead time
Calculation of fractional dead time or unavailability is restricted here to the simple case of an un-revealed fault and a regular test period, T.

The value of the fractional dead time, μ, will depend on how reliability varies with time. In general,

$$\mu = \frac{1}{T} \int_0^T F(t) \, dt$$

where, as previously, $F(t)$ is the unreliability.

For the particular case of exponential decay,

$$\mu = \frac{1}{T} \int_0^T (1 - e^{-\lambda t}) \, dt$$

$$= 1 - \frac{1}{\lambda T} (1 - e^{-\lambda T})$$

With the further restriction that $\lambda T \ll 1$. $\mu = \lambda T/2 =$ the mean failure rate $\times$ half the test period. This is the expression used in Chapter 3 if we put $\lambda = f$.

Many other situations can be dealt with relatively simply, for example, where parallel and standby redundancy are present with one or more simultaneous repair, and with revealed and unrevealed faults. Our example is adequate to illustrate the principle.

Note that the condition $lT \ll 1$ is fairly often valid in practical situations. Somewhat less frequently valid is the basic assumption of an exponential failure density function.

Index

A320 avionics system, 113
absolute probability judgement (APJ), 61, 77
absorbed dose, 138
acceptable risk, 209, 212
accident and incident data, 78
accident prevention, 68
accident sequence evaluation programme (ASEP), 64
accuracy of predictions, 166
Advisory Committee on Major Hazards (ACMH), 215
Advisory Committee on Toxic Substances (ACTS), 133
Advisory Group for Aerospace Research and Development (AGARD), 77
AEA Technology Data Centre, 73
Airbus A310, 113
Airbus A320, 113
aircraft, 20, 41, 42
aircraft safety, 31
American Conference of Governmental Industrial Hygienists (ACGIH), 133
analysis of work activities, 194
antagonism, 129
artificial intelligence (AI), 115
audits, 241
awareness, knowledge, attitudes and skills (AKAS), 248

'bathtub' curve, 17
behavioural psychology, 62
benchmark exercises, 168

best practicable environmental option (BPEO), 224
beta ratio, 27
Bhopal, 3, 271
Birnbaum–Saunders distributions, 18
blast protection, 75
Boeing 777, 114
boiling liquid expanding vapour explosion (BLEVE), 123
burning in, 17

cancer, 125, 128, 132, 137, 140, 146, 156
Canvey Island, 160, 171, 216
carcinogens, 128, 149
cascade failure, 26, 27
chemical reactor, 29
chemical safety data, 73
chemical substances, 118
chemical systems, 37
Chernobyl, 3, 122, 139, 143, 150, 260
cognitive processing, 83, 86
Comet aircraft tragedies, 36
common mode failure, 26, 29, 32, 41
complete health and safety evaluation (CHASE), 241
completeness uncertainties, 164
complex configurations, 25
component reliabilities, 166
component reliability data, 72
computer aided design (CAD), 107

computer numerically
   controlled (CNC), 107
computer-based systems, 58
containment failure, 120
control limits (CLs), 133
Control of Industrial Major
   Accident Hazard Regulations,
   1984 (CIMAH, 1984)., 134
Control of Substances
   Hazardous to Health
   Regulations 1988 (COSHH,
   1988), 74

decision sequence diagram, 237
deflagration, 38
Department of the
   Environment, 222
dispersion, 120
diversity, 28, 29, 31, 108, 109,
   113, 114
dose equivalent, 138
Dow or Mond Index, 75

ecotoxicity, 130
electrical interference, 108
electricity supply, 22
embrittlement, 36
emergency landing of a Boeing
   737, 277
environmental assessment, 226
environmental damage, 151
environmental impact
   statements, 213
environmental movement, 216
Environmental Protection
   Agency, 142, 222
epidemiology, 131
ethylene oxide plant, 30
European inventory of existing
   commercial chemical
   substances (EINECS), 75
event tree, 63
event tree analysis (ETA), 44,
   165, 196
expert systems, 65
explosion, 38, 122

explosiveness, 73
exponential distribution, 282
exponential lifetime, 13
   distribution, 15

fail safe design, 23
failure density function, 14, 15,
   281
failure modes, 16, 108, 115
failure modes and effects
   analysis (FMEA), 42, 165, 196
fatigue failure, 35
fault management, 60
fault tree analysis (FTA), 46, 59,
   165, 171
fire, 122
fire hazards, 37
fire triangle, 122
'Fitts List', 97
flammability, 73
Flixborough, 266
'fly-by-wire' (FBW), 113
FN curve, 159
fractional dead time, 24, 29, 283

gamma distribution, 18
generic error modelling system
   (GEMS), 61
gray (Gy), 138

harm, 145
hazard, 5, 68, 145, 194
hazard analysis, 34, 41
hazard and operability
   study(HAZOP), 40, 196
hazard identification, 194
hazard rate, 281
'hazardous event', 195
hazardous situations, 195
hazardous waste, 220
hereditary, 137
heuristics, 178
human behaviour, 50
human cognitive reliability
   method (HCR), 62
human error, 50, 52, 57, 58

assessment and reduction
  technique (HEART), 61
databases, 76
probability (HEP), 61
Human Error Reliability
  Analysis eXpert, 65
human factors, 2
human factors/engineering
  database, 77
human performance data, 52,
  65
human reliability (HR) data, 4,
  75
human reliability analysis, 3
human reliability assessors
  guide, 61, 64

individual risk, 158
influence diagram approach
  (IDA), 61
injuries, 147
injury, 145, 154
instrument landing system
  (ILS), 197
International Commission on
  Radiological Protection, 141
international safety rating
  system (ISRS), 241
ionizing radiation, 40, 137

job safety analysis, 196

land use, 215, 219
'latent' accidents, 51
LC_{50}, 126, 130
LD_{50}s, 130
local planning authorities
  (LPAs), 214
lognormal distribution, 18
Lusser's law of reliability, 12, 26

maintenance, 10
management of errors, 251
Management of Health and
  Safety at Work Regulations
  1992 (MHSWR), 192

maximum exposure limits
  (MELs), 133
mean time between failures, 20
mean time to failure, 20
mechanical failures, 9
mechanical systems, 35
memory, 86
men and machines, 97
Mexico City, 269
modelling uncertainties, 165
modular design, 10
mortality, 154
Mossmorran action group, 216
mutagens, 129, 149

NASA shuttle 'Challenger', 257
National Institute of
  Occupational Safety and
  Health (NIOSH), 149
National Radiological Protection
  Board, 140
National Research Council, 205,
  206
NIMBY: Not In My Back Yard,
  216
normal distribution, 282
normal lifetime distribution, 13
nuclear fuel, 46
nuclear industry, 2
nuclear radiation, 36
nuclear reactor shutdown, 28
nuclear reactor systems, 28
Nuclear Regulatory
  Commission, 142

occupation exposure standards
  (OESs), 38
occupational diseases, 146, 148
occupational exposure limits
  (OELs), 133
occupational exposure
  standards (OESs), 133
operator errors, 50, 59
organizations, 51
  safety and reliability, 101

paired comparison (PC), 61
parallel redundancy, 21, 29, 31
parallel reliabilities, 12
parameter value uncertainties, 165
perception of risk, 177, 182, 187
performance monitoring, 234, 241
person-machine interface (PMI), 96
Piper Alpha, 273
population stereotypes, 99
post-traumatic stress disorder, 150
potentiation, 129
pressure switch, 42
pressure vessels, 35, 41
probabilistic risk assessment (PRA), 5, 163, 204, 212
programmable electronic system (PES), 107
prototypes, 9
psychological disorders, 150

quantified risk analysis (QRA), 5, 163, 175, 204

radiation doses to the public, 139
Radiation Effects Research Foundation, 140
radiation weighting factor, 138
recommended limits (RLs), 133
redundancy, 108, 113, 114
reliability, 2, 8, 11, 20
Reliability Analysis Centre, 73
repetitive strain injury (RSI), 96
rich pictures and spray diagrams, 235
risk, 157, 197
 analysis, 193
 assessment, 175
 communication, 204, 206, 208, 217, 220, 225

control strategies, 240
 estimation, 192, 193
 evaluation, 175, 176, 193
 homeostasis, 186
 management, 5, 172, 204, 206, 214
 matrix, 198, 199
robot safety, 111
robots, 111
RoSPA Five Star audit, 242
Royal Commission on Environmental Pollution, 222

safety and reliability, 1, 2
safety culture, 243
Scientific Group on the Methodologies for the Safety Evaluation of Chemicals (SGOMSEC), 128
selection and training, 94, 102
sensory defects, 84
series reliabilities, 11
set maximum exposure limits (MELs), 38
sieverts (Sv.), 138
Sizewell B, 171, 212
skill, rule and knowledge (SRK) models, 59
societal risk, 158
socio-technical systems, 81, 91
software engineering, 109
somatic, 137
Standard Industry Classification (SIC), 155
standardized mortality ratios (SMRs), 154
standby redundancy, 22
static analysis, 110
stress, 91, 149
stress or fatigue, 56
subjective evaluation of risk, 177
success likelihood index method (SLIM), 61

synergism, 129
systems models of accidents, 256

technique for human error rate
    prediction (THERP), 61, 76
tecnica empirica stima operatori
    (TESEO), 61
teratogens, 129, 149
Three Mile Island, 51, 150, 171
threshold limit values (TLVs),
    39, 133
time variation of reliability, 13
tolerable and acceptable risks,
    188
toxic chemicals, 124
toxicity, 38, 73, 124
training, 101, 102, 103, 245, 246
transport systems, 51, 67

unconfined vapour cloud
    explosion (UVCE), 123
United Nations Scientific
    Committee on the Effects of
    Atomic Radiation
    (UNSCEAR), 142
unreliability, 11

visual visplay units (VDUs), 96
voting procedures, 24

WASH-1400 reactor safety
    study, 171
Weibull distribution, 17
Windscale incident, 28
x-rays, 137

yieldpoint, 35